OFFSHORING IT SERVICES:

Offshoring Management,
2nd revised edition

K MOHAN BABU

OFFSHORING IT SERVICES

Focus on Offshoring Management, 2nd revised edition

Printed in the United States of America
First Printing, June 2012

ISBN-13: 978-0615677118
ISBN-10: 0615677118

Mohan Babu K
PO Box 49441
Greensboro - NC 27419
http://www.offshoringmanagement.com

Contents

SECTION I: OFFSHORE MANAGEMENT CONTEXT

Introduction and Structure and content of this book ......

Chapter 1: Trends in Offshore IT Outsourcing — 3

Tracing the history of Offshore IT Outsourcing — 5

Offshore outsourcing: The Strategic imperative — 12

Building the Offshoring Strategy — 16

Offshore Outsourcing: Interview and perspectives — 19

Outsourcing Bandwagon: Corporations are not the only drivers — 24

Chapter 2: Planning Offshoring — 29

Outsourcing: Two Sides of the Coin — 33

Offshoring Models — 37

Risks of Offshoring — 50

Selecting the Offshoring Model — 56

SECTION II: OFFSHORING MANAGEMENT FRAMEWORK

Chapter 3: Framework for Managing Global IT Projects — 61

Offshoring Management Framework [OMF] — 63

Governance Layer — 68

Service Level Agreement [SLA] — *73*

Transitioning Offshoring — *75*

Managing Offshoring Programs — *78*

Chapter 4: Offshoring: The IT Management Context — 85

The Management Layer — *86*

Global Project Management — *92*

General Body of Knowledge — *95*

Organizational Practices and Tools — *98*

Experience and Knowledge — *102*

Globalization and Cultural Awareness — *104*

The Global IT Manager — *106*

Chapter 5: Project Execution Layer — 113

Planning — *115*

Controlling and Monitoring — *117*

Closing — *118*

Change Management — *120*

Quality — *123*

Customer-Vendor Relationship Focus — *124*

The Offshoring Sweet Spot: Project Execution — *126*

Chapter 6: Project Execution Layer: Application Development — 135

Application Development and Software Engineering — *137*

The Development Life Cycle — *138*

Managing the Application Life Cycle — *145*

Offshoring Application Development — *150*

Chapter 7: Project Execution Layer: Maintenance — 155

Maintenance Life Cycle — *161*

Offshore Management of Application Maintenance — *165*

Challenges of Managing Maintenance — *170*

Chapter 8: Communication Layer — 175

Communication Context — *176*

The Communication Layer — *180*

Tools and Technologies of Communication — *183*

SECTION III: GLOBAL ENVIRONMENT

Chapter 9: Managing Globalized Workforce — 201

Cultural Aspects of Offshoring — *206*

Managing Technical Aspects — *214*

Human Aspects — *220*

Chapter 10: External landscape and Offshoring Management — 229

Technology Landscape — *233*

Knowledge Management — *236*

Emerging techniques from Special Interest Groups — *241*

Globalization and Economic Environment — *243*

Digital Security and Offshoring — *245*

Staying the Course — *251*

Conclusion — *254*

Chapter 11: An Offshoring viewpoint on Pre-sales — 257

People Supporting pre-sales — *259*

Responding to pre-sales Requests — *261*

Best Practices in Information Technology pre-sales — *263*

**Chapter 12: Enterprise Architects Enabling
Strategic Global Sourcing — 265**

Enterprise Architects and Sourcing — *266*

Managing and addressing EA challenges in a sourcing context"
(capitalize EA) — *268*

Appendix A — 279

Appendix B — 289

Index — 293

Preface

As a technocrat who has spent over fifteen years in the dynamic field of Information Technology management, I continue to learn the intricacies of what really makes projects tick, executives sponsor technology initiatives, and users accept systems developed by technologists. During the span of my career, I have donned many hats including putting time in the trenches coding in application development languages, working with users in defining system requirements, designing and architecting solutions and managing fellow techies working towards the elusive *delivery excellence*. In my eclectic experiences, I have come to appreciate the fact that while the primary focus of technologists is to work on myriad technologies, the business problems that we attempt to solve are the raison d'etre of our existence. While each project, location, client and domain has offered its unique learning, there is a common thread running through: focus on delivery under pressure, on time, with limited resources; essentially the need to do more with less.

Lowering the total cost of IT and streamlining the development life cycle is an eternal quest in the world of business. Not surprisingly, offshoring of Information Technology has gained momentum in recent times with promises of streamlining the application lifecycle, and drawing on efficiencies of time and space, while reducing costs. Senior business executives and leaders are beginning to buy in into offshoring as a strategic tool in their arsenal; and IT managers are being expected to deliver on the promised benefits.

IT managers and technologists at both ends of the spectrum -- at outsourcing organizations and at service delivery firms, onsite and offshore -- are increasingly being expected to take on responsibilities of managing *offshored* projects. Though there continues to be some mysticism about offshored IT projects, managing them involves regular dynamics of technology and people, albeit with the added dimensions of managing cross-cultural and geographic challenges. I have been researching offshoring models including *Global Delivery Models* and *Offshoring Delivery Models* used by software service vendors -- including the GDM of my employer, Infosys -- and have been attempting to distill the commonality in the Body of Knowledge that exists among these vendors. It should be noted that although there is some osmosis and ideation based on my experiences at Infosys, one of the largest offshore outsourcing firms, I have consciously attempted to avoid direct references to their proprietary processes.

In this book we will examine some of the emerging trends and practices of managing offshored projects, explore some practical ideas, and introduce the Offshoring Management Framework, a vendor-neutral way of approaching offshoring strategies for execution and management. The Framework may be adopted by managers at 'both ends' of the offshoring spectrum: at outsourcing organizations and at service delivery firms, onsite and offshore. Large IT service companies have already begun to tailor generic processes to extend their offshoring capabilities, some have also begun to train their mid-level managers on offshoring. The thrust is on enabling their managers to be comfortable in managing global delivery projects. On the other side of the spectrum, managers and executives at sourcing organizations are equipping themselves to understand the intricacies of offshoring, managing vendors and teams distributed across the globe. This book will attempt to examine some of the emerging trends in managing offshored projects and add to the existing and emerging knowledge base.

Structure and Content of This Book

In this book we look at some of the key facets of offshoring and focus primarily on 'how to' facilitate and succeed in offshored initiatives. It must be noted that we will not debate the merits of offshoring or justify its need. The book is divided into three main sections; the first section analyzes the offshoring context, landscape and environment. The second section introduces the Offshoring Management Framework and the intricacies involved in implementing and managing outsourced IT. The last section examines aspects of the global environment.

The *first* chapter analyzes some of the emerging trends in Offshore IT outsourcing, along with a brief look at the history of offshoring, the strategic imperatives and aspects of building an Offshoring Strategy. The *second* chapter examines the salient aspects of offshoring management and planning including a brief analysis of offshoring risks; the chapter also looks at some of the popular offshoring models.

The *third* chapter introduces a Framework for managing global IT projects, that we shall term the *Offshoring Management Framework [OMF]*. The chapter details the Governance Layer with aspects of program management and management of Service Level Agreement [SLA]. Managing the transition from existing IT infrastructure and processes to offshore teams is an emerging area of focus, and in the chapter we will look at how organizations transition from an *observer* to a *strong* offshoring player.

The success of technology outsourcing lies in the consistent and predictable delivery of projects and products; and may include reliable maintenance of applications. After the sourcing strategy has been defined and piloted, operational aspects including managing delivery of projects and coordination between offshore and onsite teams acquires greater significance. The *fourth* chapter sets the management context and delves into aspects of Global Project Management.

The Project Execution Layer is the most significant layer of the Framework since it addresses the tactical aspects pertaining to individual projects and work executed at the lowest level of granularity of offshoring, involving collaboration between onsite and offshore teams. The fifth, sixth and seventh chapters delve into aspects of execution of offshored projects with an emphasis on the aspects of Application Development and Maintenance.

Managing globally distributed teams and onsite-offshore coordination requires clear, unambiguous communication between individuals and teams, and communication in a business context continues to be a significant area of focus and concern. The eighth chapter describes aspects of the Communication Layer and includes a discussion on Tools and Technologies of Communication.

Software development is an intellectual activity that cannot be accomplished without groups of skilled and talented people synchronizing their efforts towards a common goal. Offshoring necessitates interaction, communication and networking with partners, vendors, suppliers, outsourcers and others from around the world. The key challenge is to manage the dynamics of global workforce, also called 'geographically distributed teams' or 'virtual teams.' The ninth chapter is about Managing Globalized Workforce and includes discussions on cultural aspects of offshoring.

Aspects of external landscape including changes in technologies, lobbying by special interest groups, economic upheavals or governmental policies can directly or indirectly influence offshoring initiatives. The book concludes with an analysis of some of the key aspects pertaining to external landscape and Offshoring Management including the Technology Landscape, Knowledge Management, Globalization and the Economic Environment.

In the revised edition of the book, I have added sections on pre sales and Enterprise Architecture as they pertain to the business of offshoring.

Mohan Babu K
North Carolina, August 2012

CHAPTER 1

Trends in Offshore
IT Outsourcing

- 💻 Tracing the History of Offshore IT Outsourcing
- 💻 Offshore Outsourcing: The Strategic Imperative
- 💻 Building the Offshoring Strategy
- 💻 Offshore Outsourcing: Interview and Perspectives
- 💻 Outsourcing Bandwagon: Corporations are not the only drivers

---

Offshore outsourcing as a strategy for management of Information Technology (IT) development and maintenance has gained momentum during recent times. A trend that began as a tactic to move low-end IT work to offshore locations to cut costs of business computing has now entered the realm of mainstream decision making, which some call a *mega trend*[1]. The increasing focus on offshoring is prompting business leaders and management gurus to examine successful practices of globalization.

Offshoring of IT is not an isolated trend but part of a bigger shift towards the globalization of business processes. Echoing this, Gregory Millman[2] says, *'As outsourcing has moved beyond IT to include a broader range of business processes, it has evolved from a short term tactic into a long term strategic play.'* Western companies began the process of outsourcing decades ago when they began to manufacture

3

physical goods in low cost centers such as Taiwan. Offshoring IT systems and business process followed a similar trend and was aided by faster and cheaper communication technologies. We shall define offshoring as:

> *A strategy of relocating business processes, services and work to overseas locations, where it makes most business sense, by capitalizing on the global skill pool, advances in communication technologies and the benefits of cost arbitrage.*

The above definition draws on the raison d'etre of offshoring—the cost arbitrage driving business strategies. We will use this definition of offshoring (used as a noun) in this text, though a few other definitions[3] also exist. The terms offshoring, outsourcing and offshore outsourcing are also used interchangeably. Although this definition of offshoring encompasses business processes and services in the broader business context, we will primarily focus on IT offshoring. Some of the major changes in the business climate bringing a renewed focus on global-ization of teams include:

Geopolitical factors of globalization: Globalization is no longer just a buzzword; it is a business paradigm that most enterprises are actively embracing. Interestingly it is not just the mega corporations but even micro firms in the West that are actively contemplating outsourcing[4].

Shift towards project based organizations: Functional or pro-cess based organizations were the norm in the business world until recently; strategic operations were controlled by managing the different functional areas of business like Finance, Production, Manufacturing, Sales and Marketing etc., operating as independent silos. There is a shift from this model towards project based organi-zational structures where teams of functional experts get together for project initiatives after which they move on to other teams. This trend has also been accentuated by the increasing adoption of enterprise class software applications (ERP, CRM etc.) that cut across functional areas and automate most of the mundane tasks. The maintenance of such enterprise applications is increasingly being viewed as a candidate for outsourcing.

Cross-functional teams: Cross-functional teams are an extension of the shift towards project-based organizations. Teams of individuals from different functional areas are assembled together for specific initiatives. Such teams may have to be tightly managed because of the diverging goals, skills and views that they bring to the table. Such cross-functional teams are increasingly becoming global as managers find that they can capitalize on the availability of niche skills from a global pool.

Self managed teams: Along with the shift towards project based structures and adoption of cross-functional teams, organizations are also beginning to experiment with self-managed team structures where the role of a manager involves co-ordinating and orchestrating the efforts of the teams rather than *managing* the activities.

Some of the trends highlighted above are impacting the global business landscape and leading business executives to formulate globalization strategies. This also impacts the way IT systems and projects are managed since offshoring of technology development and maintenance is among the leading trends. This increased shift of application development towards offshore centers is leading to a demand for global managers, especially for those with a strong technical background coupled with multi-cultural skills and grounding in the basics of project management.

TRACING THE HISTORY OF OFFSHORE IT OUTSOURCING

The software industry has always had traces of globalization. The trend goes back to the nineteen fifties and sixties when software engineers and programmers across Europe and North America collaborated on research and development. Interestingly, bridging of the geographic divide was the genesis of the Internet. Outsourcing of IT work to system integrators and consolidators has been in vogue for a little over two decades, though such outsourcing was handled by domestic vendors. It is only in the past few years that offshoring has taken off in a big way. The trends in outsourcing and globalization, that were developing in parallel, merged with the dynamic changes in the way organizations began to look at IT and leverage its benefits leading to the 'boom' in offshoring.

To trace the history of offshore outsourcing, we should begin by looking at the application development and maintenance trends in the west that have gone through some very dynamic changes during the past decade. During the nineties, two successive events irreversibly changed the management view of Information Systems. First was the (then) impending Y2K threat that brought renewed management focus on legacy systems that were in production for years, in some cases even decades. The second mega trend witnessed in the IT and business world was the advent of Internet and E-Commerce. Business leaders also began to view E-Commerce as a serious threat to their existing way of functioning. Management gurus and IT leaders began to implore their corporate brethren not to miss the boat: "Internet or perish" became the mantra to live by. In many cases, large organizations that were spending millions of dollars on upgrading their IT systems to be Y2K compliant were, in parallel, trying to articulate their internet and E-Commerce strategies.

The sudden surge in requirement for IT professionals fueled by both the Y2K and E-commerce booms caught IT departments and their vendors, the software services companies, off-guard. Suddenly everyone began to scramble for resources to meet the demand. Unable to find sufficient talent in local economies, companies and software vendors went overseas. Employers and consulting organizations began to 'import' skilled professionals in large numbers. Restrictive immigration policies were one of the major bottlenecks in the western economies. The corporate world had to lobby the governments in the US, UK, Canada and else-where in the West to increase the number of visas issued to foreign workers. Hundreds of thousands of foreign IT professionals—nearly half of them from India—entered the global marketplace. America, one of the largest markets for software services, enacted a law in 1998 to increase the annual H1-B quota from 65,000 to 115,000 per year.

The IT boom in the West during the nineties translated to a surge of young talent moving into the industry in India, China, Ireland, the Philippines and other 'talent supplying' countries. *Body Shops*, a euphemism for staff supplementation firms, sprung up in several cities in these supplier nations. The large scale work under-taken by data-centers across the world to prepare for the Y2K threat brought to fore the enormous amounts of data and business logic residing in IT systems. Business leaders began to realize how indis-pensable the systems were. Staff supplementation firms in these countries began exploring the use of emerging communication technologies to connect to clients and began to take on small IT work from client locations to be worked on by their offshore teams. Some also invested in proprietary satellite links and dedicated communication infrastructure. In a sense, it was the genesis of offshore outsourcing as we know it today.

Offshoring of IT is just one of the trends in the movement towards internationalization and globalization of business and economies. The recent decades have seen the emergence and growth of multinational and transnational corporations that are comfortable operating in multiple geographies and countries. The use of pervasive, cheap communication technologies including web collaboration software, email and other internet technologies have only accentuated the progress towards globalization among mid-size and smaller organizations.

Figure 1.1 highlights the transition in sourcing models across the spectrum from in-house development to local sourcing and towards the offshoring that we are now experiencing.

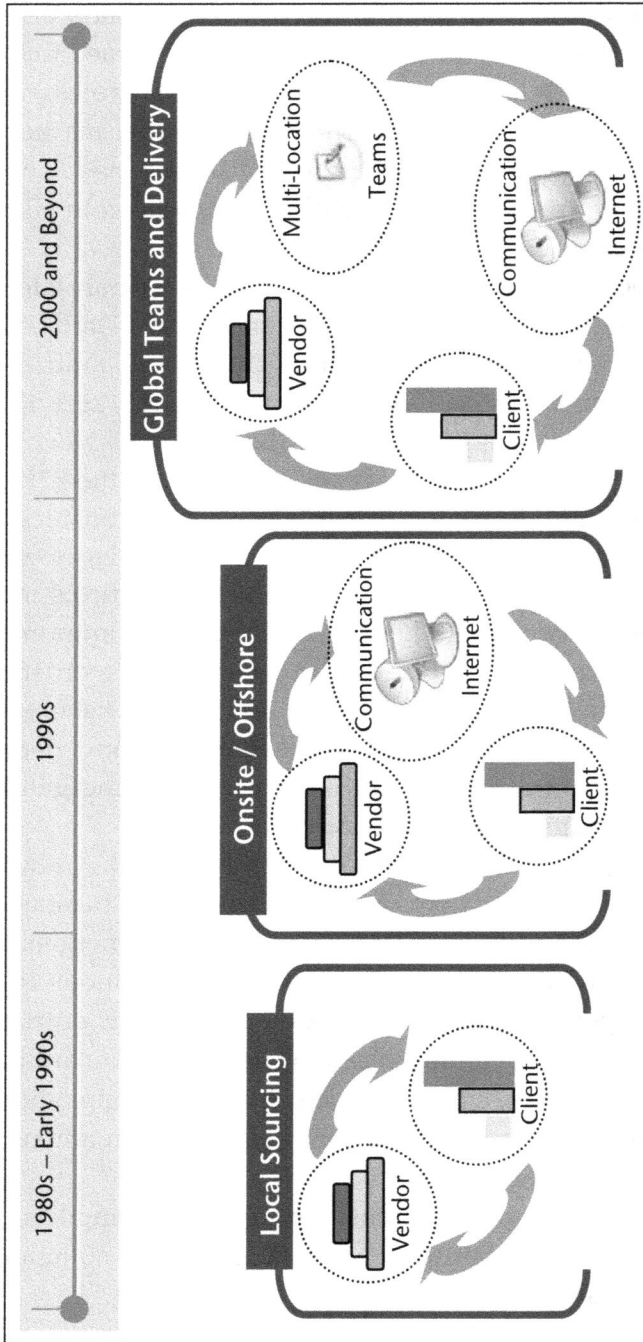

Fig. 1.1   Offshoring Trends

Local Sourcing: Pure Sourcing Model

During the nineteen eighties, organizations began outsourcing development and maintenance of technology applications to System Integrators and Service Delivery organizations. These vendors either co-located their teams at the client's data centers or executed projects out of their own offices and delivered the systems to clients. In some cases, the vendors took over the entire data center operations including systems, hardware and the staff. The features of this system were as follows:

- The staff was primarily domestic with a few international experts flown in to be a part of the teams.
- Vendor and Client teams were primarily co-located at client sites.
- Sourcing primarily involved staff-supplementation or contracting and, in a few cases, data center management.
- Challenges included managing the logistics and interaction between client and vendor teams.

Onsite/Offshore Sourcing: Mixed Sourcing Model

During the early nineties, due to the acute shortage of skills and an increasing demand for IT professionals, software service organizations began to hire foreign workers. The mode of delivery continued to be staff supplementation and the challenges included sponsoring visas and travel for technical staff to travel to client locations for project execution. Staff Supplementation companies in India, Ireland and other offshoring destinations began the attempt to move up the value chain by taking on small projects for execution offshore, primarily to augment the staffing model. This involved a bigger paradigm shift for IT managers in the west who had to formulate strategies for managing foreign workers and work performed offshore.

Some of the features of the mixed sourcing model include:

- The driver was the demand surge for foreign IT professionals to supplement domestic talent; the field of software development went global.
- The mode of delivery continued to focus on staff augmentation.
- A few large vendors also set up Global Development subsidiaries in Ireland, India and other locations.
- Challenges included cross-cultural management, logistics of visas and travel.

Global Teams and Delivery: Global Sourcing Model

Offshore outsourcing began to take off around the year 2000 when software service organizations began to offer Global Delivery Models (Refer next section). Large enterprises also began to set up their independent subsidiaries and captive IT development centers in low-cost locations overseas. Although adoption of such global delivery did not eliminate the need for technical teams to travel to client locations, it minimized it considerably. Some highlights of the trend include:

- International teams and members co-located in multiple geographic locations.
- High reliance on modern tools and technologies of communications.
- Mode of development includes Global Delivery Models and multi-site management strategies.
- Challenges include cross-cultural management and motivating teams from different cultural backgrounds and, to some extent, the logistics of visas and travel.

The trends in offshoring are leading to a stronger demand for services even as global development models continue to emerge.

Offshoring vendors and software service companies are already positioning their proprietary development models as differentiators. 'Global Delivery Model,' 'Development follows the sun,' '24 X 7 delivery,' 'Strategic Outsourcing,' and 'Offshore Outsourcing Model' are some common phrases used to describe models adopted by vendor organizations. (Ref: Appendix A—brief analysis of other sourcing models). The term Global Delivery Model alone has been used by at least a few dozen organizations. Here is a sampling of the global delivery models being adopted by some of the notable leaders in this space:

> *'The GDM (Global Delivery Model) framework works like a well-oiled mechanism where the project teams, located physically at different locations, are perfectly co-ordinated through seamless communication and clearly defined process guidelines.'* –Infosys[5]

> *'GlobeGain', Wipro's offshore outsourcing model for offshore software development guarantees cost savings upto 35% within the first 18 months with upto 35% productivity increases and 75% faster time-to-market. Robust processes, a rich portfolio of reusable frameworks, time zone advantages and seasoned project teams combine to deliver measurable business results to our offshore customers.'* –Wipro[6]

> *'IBM Strategic Outsourcing Services is the management of a companies' applications and information technology (IT) systems. Customers strategically partner with IBM to manage and operate their applications and IT systems, generally under a mutually beneficial agreement.'* –IBM[7]

The Global Delivery Models highlighted above are a small sampling from an initial online search engine scan. Most large software service organizations have begun building service offerings around aspects of offshoring to complement their suite of offerings. While service delivery organizations pitch the models as distinct and proprietary, a few building blocks that form the core are similar. The building blocks for management of offshoring will continue to be the focus of study in this book.

OFFSHORE OUTSOURCING: THE STRATEGIC IMPERATIVE

Offshore outsourcing of IT, a.k.a. offshoring, that began as a 'low cost' tactic for staff supplementation by IT managers is morphing into a strategic option available to business planners and technologists. Information Technology and Business Process Outsourcing (ITO and BPO) have emerged as mainstream trends. IT offshoring and its role in the formulation of business strategies continue to intrigue academicians and business titans alike. A recent *Cutter IT Journal*[8] affirms this trend stating 'Effective IT outsourcing requires IT managers to rethink their business objectives, their processes, their metrics, and the way they do business with vendors and suppliers.' A study published by Stern Stewart & Co[9], that analyzed 27 major outsourcing announcements and compared the performance of those companies, concluded that the companies studied outperformed the market by 5.7%. Most large software service organizations have also published their lists of 'top 5' or 'top 10' reasons for offshoring. There are three main drivers behind the trends in offshoring:

1. **Limited Talent Pool:** The migration of software workers to the West has dwindled during recent times because of tightening immigration laws. In addition, there is a demographic shift being experienced in the West as the baby boomers begin to age and retire. These and other trends in the labor-market are forcing executives to take a hard look at their staffing strategies and plan for the long term. Offshoring is one of the strategies of supplementing and invigorating the existing talent pool.

2. **Cost Pressures:** One of the key drivers behind offshore outsourcing is the lower cost of doing work. Corporations are reaching the limits of cost cutting by traditional methods and are beginning to view outsourcing as an extension of the strategy of extending IT services to their business users while lowering the total cost of operations. There is a distinct need to gain operational efficiency and leverage the scarce and expensive IT resources. There is also an element of competitive drive here; organizations and managers want to be a part of the outsourcing wave as they increasingly observe competitors lower the cost of IT operations by offshoring.

3. **Innovate Vs. Sustain:** The Innovate-versus-sustain dilemma (Figure 1.2) is a challenge managers face when reacting to innovations in the technical and business landscape; while they need to ensure that the limited resources and talent pool at their disposal manage and support the various Line of Business (LOB) applications, managers also need to ensure

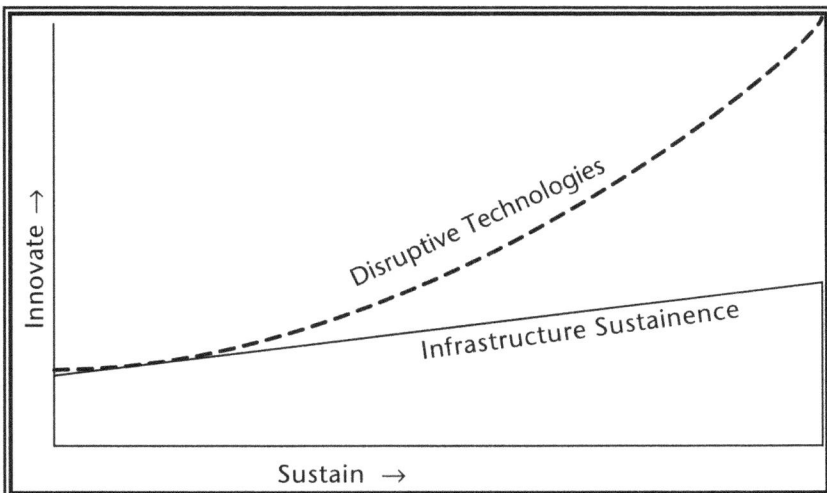

Fig. 1.2   Innovate-Vs.-Sustain

that innovations are leveraged to the extent possible. Offshoring is a key option available to take on the Innovate-versus-Sustain dilemma. By sourcing the non-core work to specialist vendors, IT managers and their teams can focus on innovations in their space and work towards capitalizing them. The cost benefit from sourcing accentuates this argument.

IT Managers have to ensure that businesses derive the maximum Return on Investment (ROI) on their IT Infrastructure while remaining agile and capitalizing on the emerging trends and technologies. A wide spectrum of businesses in verticals ranging from telecommunications to financial services are outsourcing their technology development and application maintenance either to vendors offshore, or are setting up captive subsidiaries of their own. Strong business imperatives behind offshoring are very well documented in the academic and business press, some of them are highlighted in the box 1.1 below.

Box 1.1

OFFSHORING: KEY DRIVERS

Cost pressure: Organizations continue to be under constant pressure to minimize the cost of operations and maximize efficiencies. Offshore outsourcing is increasingly becoming a powerful strategic tool to enable a better return on IT investment and organizations are beginning to weave offshoring into their corporate IT and business strategies.

Competitive pressure: As an increasing number of organizations across industry verticals are resorting to offshoring of IT to gain cost benefits, their competitors are being compelled to examine their own offshoring strategies. Executives are realizing

Box 1.1

OFFSHORING: CONTINUED...

that excluding offshoring from their strategic IT mix can be detrimental to their overall goals, especially if their competitors have already begun to reap the benefits of sourcing.

Focus on Core Competencies: The use of technology or IT systems are essential for the sustenance of most modern businesses, but only in a few cases do such systems end up being the *Core Competence*[10]. Offshoring can help business leaders focus on innovation and time-to-market issues of systems driving their core competence without having to worry about the operational aspects of outsourced systems.

Maturing IT vendors: Software vendors and application service providers in the offshore outsourcing space are maturing and improving on their capabilities. Many have been refining their outsourcing models and have successfully operationalized offshore development methodologies. By working with such vendors, organizations can reap benefits with the least risk exposure.

Maturing communication technologies: Communication tools and technologies that help bridge the onsite–offshore digital divide have emerged, and are successfully being adopted. The uses of such technologies mitigate the risks of communication and co-ordination generally associated with managing globally dispersed teams. This in turn minimizes the need for face-to-face meetings and travel, lowering the total cost.

Risk awareness: There is a greater awareness of the risks—and mitigation strategies—associated with offshore outsourcing among vendors and their clients. Companies and executives no longer consider offshoring a totally *unknown* strategy and are more confident in planning sourcing strategies that address and mitigate the known risks.

BUILDING THE OFFSHORING STRATEGY

Globalization and more rapid economic cycles are leading organizations to re-examine the way they run operations, serve their clients and compete. Companies of all genres are beginning to examine global sourcing models and are looking for ways to add them to their portfolio of strategies to manage business challenges. Decisions on globalization, as with any strategic decision making, can be very intricate and may require a lot of planning, brainstorming and benchmarking. Before embarking on a globalization strategy, an organization needs to take a holistic view of its existing operations, culture and business drivers, study the various sourcing models and adopt a roadmap that they are most comfortable with. Articulating the benefits of taking a big-picture view of offshoring, Ian Hayes[11] says '*Offshoring is not a cheaper source of resources but a different way of doing business. The sooner we and our organizations understand and adapt to that difference, the sooner offshore outsourcing will move from being a source of difficulty and anxiety to one more effective delivery option in our arsenal of IT practices.*'

The process of defining technology roadmaps, until recently the forte of IT managers, is increasingly being taken over by business executives. This makes a lot of sense since they (the business leaders) are well positioned to take decisions based on their understanding of business drivers. Technology and business roadmaps attempt to take the application portfolio from the current 'As is' to a 'To be' state. The roadmap definition exercise could be triggered by changing the business and technology landscape, and has to be integrated into a business planning exercise. Some of the key aspects of offshoring to be considered while defining an IT sourcing roadmap include:

- **Identifying candidates for offshoring:** This includes consolidating systems in the IT portfolio and planned outsourcing of non-unique LOB systems. The input for this process comes from an application portfolio analysis and offshoring roadmap.

- **Define Key Success Factors:** The expectations from offshoring projects and initiatives including expected Service Level Agreements (SLA) and other keys success factors need to be articulated by the senior management to ensure that all parties are in sync. Most business units and IT systems may already have their standard SLA templates and processes that may be extended to cover aspects of offshoring. A well defined SLA, spelling out the expected throughput, variances and other operational aspects of business, will help both the client and vendor teams retain focus.
- **Defining the Program Management Strategy:** Defining the program management strategies and the working models of individual outsourcing projects is a key step in an offshoring roadmap. Managing global projects involves a level of complexity that extends beyond project management of service delivery initiatives.
- **Operations management:** The end goal of any sourcing strategy is to optimize the steady state of operations management. After the sourcing strategy and program management aspects are defined, operational aspects of the management of the sourcing initiative and individual projects come to the fore. This also involves the definition of the working model, the 'how to' of sourcing, and will include the communication, technical and other management challenges including uncovering the on-going risks of managing in regulatory environments.
- **Benchmarking:** The success of any offshoring project or program needs to be constantly monitored both from a microlevel as well as from a strategic level. Strategies for benchmarking and tracking the health of global operations need to be planned while formulating the workings of offshoring.

IT applications ranging from maintenance of legacy systems to development of state-of-the-art systems and R&D are increasingly being outsourced to offshore locations and vendors. Offshoring involves mapping of business and IT strategies to the operational tactics and working models of offshoring that will be deployed by different business units, groups and projects. Finding common ground or the 'sweet spot' between the strategic goals of businesses and the changing technical landscape and capitalizing on the synergies is a topic business and technology leaders ponder over.

Outsourcing being undertaken by most organizations is not a short term trend or just a cost-cutting measure. While most business leaders are beginning to take a long-term view of their global sourcing strategies, there are doubts among a few if this shift towards outsourcing will continue in the long run. They fear that the outsourcing boom that we are experiencing is a "one off" kind and companies will roll back their offshoring strategies the moment the economy starts moving north. However, there is little merit to such an argument.

The cost benefits, along with other reasons that justified outsourcing may prevent its movement back. One can perhaps draw a parallel between sourcing of white-collared jobs and the move of manufacturing overseas. Shailesh Singh, an IT veteran based in Colorado, who has seen numerous swings in the industry likens the current trend in outsourcing to the shift of low-tech manufacturing of consumer goods like hosiery and sneakers out of the US about three decades ago. Needless to say, none of that manufacturing has returned back to the US. He says, "*Just like we may never see the return of manufacturing of shoes in the US, even though the manufacture is controlled in America by the likes of Nike, the trend towards IT outsourcing may be irreversible.*" Just like Nike and Wrangler remain quintessentially American by managing the bulk of overseas manufacturing from hubs in the US, Adobe, EDS, IBM and Microsoft will remain American even if the bulk of software development is done in India. On a similar vein, Hayes[11] adds, "*Like the many changes that have come before, however, outsourcing is here to stay, and we need to accept and adapt to that reality. Once the smoke, hype, and rhetoric disappear, we'll discover that outsourcing won't replace all (or even most) of our jobs,*" *but it will become a permanent part of our IT resource strategy, work assignments will be split differently, and we'll be responsible for making it succeed."*

OFFSHORE OUTSOURCING: INTERVIEW AND PERSPECTIVES

In this section, we briefly examine some of the key aspects of globalization and offshoring in an interview with Mark Kobayashi-Hillary[12].

(Question: Q, Mark: M)

Q: There are probably three main dimensions to managing global teams: a) cultural aspects, b) technical aspects and c) human aspects. Although management textbooks stress on aspects pertaining to people management and the importance of soft skills, it is perhaps overlooked in the field. What are your thoughts on this?

M: *I think you have identified the three main factors. While the textbooks often concentrate on the soft-skills required—it is very different managing a team in Bangalore to a team in Paris—there are also practical difficulties. Instant Messaging tools allow a team that is scattered across the world to be connected and facilitates chat and easy collaboration on PC software. Software companies, such as Microsoft and Oracle, have woken up to the fact that their tools are being used by global teams and are building facilities into the tools that help the user to work with others remotely. Ultimately though, the manager of an international team needs to meet them and have some face time, regardless of the technology. The manager should be aware of the different drivers and aspirations in each location so the team can be managed appropriately.*

Q: If selling offshoring and its benefits to clients is hard, executing offshored projects is harder. Any comments?

M: *Selling the offshore model is often hard because it may be considered to be a high risk strategy, therefore the potential downside outweighs the risk of doing it. I personally feel that there are five key areas for an* *offshoring project and these need to be examined before the process begins and managed closely if a decision is taken to go ahead:*

1. **Search costs**—*All the questions about whether to create an off-shore subsidiary, whether to partner with an outsourcing vendor, whether to go to India or the Philippines.... All those questions can take up an enormous amount of management research time and need to be resolved before you can consider the project.*

2. **Project management**—*Managing an outsourcing program is very different to regular line management. Don't pull project managers from one team (especially IT) and expect them to manager vendor relationships. You need to think about who the key people are and what is their role.*

3. **Culture**—*This is applicable where the contract is taking your organization to a distant location, such as a US firm outsourcing to India. The management needs to be aware of how to work in the new environment and the new hires in the offshore location need a good appreciation of the company culture.*

4. **Change**—*Outsourcing is usually a part of a bigger process of Business Process Reengineering, or just improving the way things are done to make them more efficient. However, change is hard as most employees don't like it and will resist, so in addition to managing a knowledge transfer process for the outsourcing deal you have to ensure that the new working methods are adopted at the onshore location.*

5. **PR**—*You need a good internal and external communication strategy. Several large companies have found themselves suffering very poor newspaper headlines because of outsourcing programs and those same programs can be almost impossible to achieve if the internal teams do not believe in them. Make sure you can communicate the benefits from top to bottom.*

Q: Team building and management of team dynamics is hard enough without globalization in the equation. Do you agree that cross-cultural aspects of globalization can be an invigorating challenge for any manager?

M: *It is the greatest experience any manager can have. When you can see your team working well together and succeeding and you take a step back and think, I have a team here spread across 8 countries, comprising 15 nationalities, and they are all working together to achieve the same objective, it's a nice feeling to be marshalling that kind of team together. Low cost telecommunications and the Internet have made the business world smaller so this is the next challenge for managers everywhere.*

Q: You have dwelt with subtle issues of offshoring such as culture and relationship management extensively in your book. Can you summarize the essence for the readers here?

M: *When I started working in India, I was managing a technology team for a European investment bank. I was a foreigner trying to manage people there and although I had previously worked with many Indians in Singapore I still felt as if I was learning something new everyday about how to deal with the situation. I started looking around at bookstores and on the Internet, to see if any managers had written about this. Considering that the newspapers were running stories about India and outsourcing almost every day, I was surprised to find that there were no books on the subject. The nearest I could find was academic tomes on the efficiencies of outsourcing or how to make IT outsourcing work—nothing on what it is like for the poor British guy sent to make it work in India.*

So I decided to make some notes and put a book together myself. It would cover all the points I found it useful to learn about India, such as some intro-duction to the culture and politics of the country as well as a primer on the major companies offering sourcing and the cities where the work is taking place. I wanted it to be a good guide for the first-time manager to read on the plane from JFK to Delhi. It covers some of the more subtle points about managing in India, because I had to go through those processes first-hand and I felt it would be useful for other foreign managers to learn about the pitfalls from my book, rather than making the same mistakes I did.

Q: Managing individuals, their goals and aspirations is one of the key management imperatives. What would your advice to a project manager managing a cross-cultural team be?

M: *It is actually quite similar wherever you go. Any manager must distinguish between the aspirations of the individual and what the manager requires of the person. So it is best to always agree on a twin-track development program, where you are training the individual specifically with the next promotion or role in sight as well as encouraging some other skills. A good example is how my Indian team in Singapore were taking lessons in Mandarin, not a direct requirement of the job, but useful for their future job prospects.*

A key difference in India is the requirement to keep on learning. People insist on learning from a job and this can be dealt with through a process of delivering a smaller amount of on-site training far more often than the more common off-site training typical in Europe.

Q: Are the 'basics' of human resource management still significant in the global human resource management context? (By basics, I mean Maslow's Hierarchy of Human Needs, Herzberg's study of what motivates people to work and other fundamental theories that managers generally study.)

M: *Yes, I think the basic psychological maps of what drives people to work are still valid when applied across cultures. All the teams I have managed in Japan, India, Singapore, France, the UK and US have all been driven by a desire to provide for their family, to learn and to advance their career. There are subtle variations in the importance of different aspects of the job, but fundamentally there is no difference.*

Q: Managing the transition of outsourced projects is among the most challenging tasks faced by a project manager. How does one go about mitigating the risks of such a transition?

M: *I mentioned my five key points on this earlier. I would stress that before even considering where to go and who to partner with, the management team must consider what is to be outsourced and why. Almost all outsourcing projects are driven by cost savings at present, however,* some companies are starting to look beyond this to the value that can be created through outsourcing. Generally this is from tapping into larger pools of good resources and allowing the company to focus on its core competence—doing what it does best and outsourcing the rest. By considering what should be outsourced and why the company should work in this way and justifying the potential benefits against the risk of a project failure, the company can begin to formulate a plan of action.

Q: Software development is an intellectual activity that cannot be accomplished without groups of skilled and talented people synchronizing their efforts towards a common goal. Increasingly, software development and other high-end intellectual activities are being offshored. How can individuals prepare themselves to benefit from such globalization?

M: *Some high-end intellectual activity will be offshored, because it can be performed in any location by skilled individuals. Those who remain in the high-cost locations, such as the US, will not lose all these jobs and they should benefit from being able to partner with other locations.*

The challenge is in presenting a blend of skills to the marketplace. Where a computer programmer skilled in C++ or Java might have previously been a well rewarded employee, these are now commodity skills that can be sourced on the global labor market. The present generation of C++ and Java programmers needs to offer skills in programming and financial services or retail distribution or oil production—those business skills remain local and are required by the local employers.

Individuals should benefit from this process as it matures. It is not a race for the bottom as jobs vanish. Many more jobs are lost as companies move from one state to another in the US than are lost to offshoring. As the economy continues to grow and the number of people in the workforce declines there will be a need to offshore some work and to promote immigration. The economy in countries such as the US will respond to this entire shift towards a global labor force by ensuring that its people offer a higher value-added service than that available from overseas.

OUTSOURCING BANDWAGON: CORPORATIONS ARE NOT THE ONLY DRIVERS

Management gurus and business leaders have been emphasizing the benefits of outsourcing and the need for corporations to jump the bandwagon amidst opposition from those wishing to protect their local economies, fearing job losses. The flip side of this (sometimes) heated debate is that the message on outsourcing has not gone unnoticed by individuals taking the lead by preparing to be a part of the global marketplace. A mail from a reader of my column (Box 1.2) was intriguing in the context of a globalized workplace and culture. Debbie's query is perhaps a consequence of the multiculturalism that organizations are increasingly grappling with.

Box 1.2

MAIL QUERY (GLOBALIZATION: EAST MEETS WEST?)

Dear Mohan,

In an Internet search for materials and content for a human relations project at school, I came across a link to your page on the business trends for IT professionals. It is my hope that you may be able to assist me with some information or possibly point me to some links on the Web that might help with my project.

    To give you some background, I am a CIS major at a community college in Texas. As part of our degree plan or track, we are required to take a course in Human Relations. The course covers communications skills, conflict management, teamwork and team building skills, etc. Our final for the semester is to do a presentation about "Managing across cultures"—meaning we are to study another culture that we would most likely work closely with in our field once we

Box 1.2

MAIL QUERY   CONTINUED...

graduate. As either a colleague, manager, or a supervisor, what steps might we take to identify the important aspects of each culture in our department or within our team? Then determine which could be applied to the workplace and modify the work environment so that each team member of various cultures feels comfortable, fosters teamwork, and makes for a more productive workplace.

With the shift in the IT world, more and more IT Professionals from India and the US are working and collaborating on projects. As I will be graduating soon, I'd like to know what resources might be available to help someone learn more about the Indian culture and what might an IT professional from India expect from other IT professionals, especially western/North American IT professionals. I genuinely would like to know and understand for both myself and for completing my project.

Would you have some time to give me some feedback or point to some materials or information?

Sincerely,
Debbie X *(Name changed to protect identity)*

People across the globe embarking on new careers are realizing the significance of cross-cultural workforce dynamics and are making conscious efforts to gain a heads up. The fact that jobs and projects are being executed across borders also means that individuals need to learn new skills of multiculturalism. As the focus on software development across geographic boundaries increases, it is expected that there will be a renewed need for individuals to acquire skills and knowledge required to manage global projects. Universities in the west are already supplementing regular Project

Management training with courses tailored towards management of offshore development. Washington University in St. Louis (Ref: Appendix A) is perhaps attempting to address the needs of Debbie along with the demand from corporate managers. In the rest of this book, we will continue to focus on aspects of globalization, with an emphasis on management of technology projects.

NOTES

1. *'It is to anticipate the next mega trend, and yet not get distracted by the noise of a hundred talking heads.'* [Annual Report of Infosys Technologies; Year 2003–04]

2. 'Going Beyond Commodity Outsourcing' [Gregory J. Millman, *Financial Executive*, September 03, Special Section]

3 Definitions of offshoring include:

   ❑ *What is Offshoring?* There is no official definition of the term "offshoring," but it has come to mean the actions of American firms in relocating some part of their domestic operations to a foreign country, including, for example, automobile firms switching purchases of auto parts from domestic plants to Mexico; computer or software firms transferring some of their programming operations to India; or financial firms relocating major parts of their record-keeping activities to one of the Caribbean countries. —Brookings Institution [Policy Brief #136, http://www.brookings.edu/comm/policy-briefs/pb136.htm]

   ❑ Offshoring can be defined as relocation of business processes (including production/manufacturing) to a lower cost location, usually overseas. —Guru.net [http://www.guru.net]

   ❑ Offshoring, can be defined as relocation of business processes (including production/manufacturing) to a lower cost location. —Investor Dictionary.com [http://www.investordictionary.com/definition/Offshoring.aspx]

   ❑ Noun: offshoring—The relocation of business activity to a location in another country with lower costs "Nearly half of Europe's top firms plan more offshoring in coming years"—WordWeb Online [http://www.wordwebonline.com/en/OFFSHORING]

4. Hearing 'I Work Cheap' From Across the Globe [Lee Gomes, *The Wall Street Journal*, June 13, 2002]

5. Infosys [http://www.infosys.com]

6. Wipro [http://www.wipro.com/]

7. IBM [http://www-5.ibm.com/services/be/so/ or *http://www-ibm.com*]

8. Cover Page [*Cutter IT Journal*, October 2004]

9. A study on EVA published by Stern Stewart & Co, 2000 *http://www.sternstewart.com/*

10. Hamel, Gary & Prahalad, CK, The Core Competence of the Corporation. [*Harvard Business Review*, May–June, 1990]

11. Breaking the Cycle: Rethinking your approach to Offshore Outsourcing [*Cutter IT Journal*, October 2004, Ian S. Hayes]

12. Mark Kobayashi-Hillary is an Executive Director of Commonwealth Business Council Technologies Ltd and fx Auctions plc. He is the author of *Outsourcing to India: The Offshore Advantage*. In the book, he offers his views on the trend to outsource to India, describing the reasons why a business should utilize India as an offshore outsourcing destination and the steps needed to find and work with a local partner. He has dwelt extensively on the softer side of cultural and cross-cultural issues pertaining to offshoring, with a focus on India.

CHAPTER 2

Planning Offshoring

- Outsourcing: Two Sides of the Coin
- Offshoring Models
- Joint Ventures
- Risks of Offshoring
- Selecting the Offshoring Model

In the previous chapter, we looked at some of the trends in offshoring along with an analysis of the strategic imperatives involved. Senior management and business leaders pursue offshoring for various reasons, but the key driver may be to ensure that their organization remains agile and competitive and is able to capitalize on the benefits of sourcing. The strategic decision of outsourcing to offshore locations includes a transition management plan. The actual distribution of tasks, activities and projects across onsite and offshore locations vary depending on the specific needs of the projects and initiatives. Some applications, programs and projects may be more *offshorable* to teams across the globe than others. Similarly, the risk tolerance and overall globalization strategies of organizations may vary; some companies, because of their business models, size and geographic spread, may be more willing to explore offshoring strategies.

Building an offshoring strategy involves developing and implementing a comprehensive technology and management transition

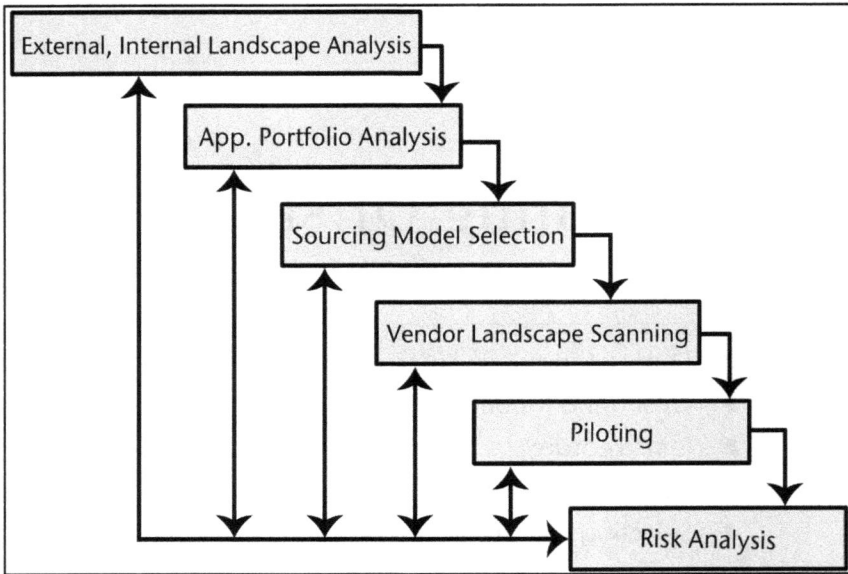

Fig. 2.1   Iterative Waterfall Model of Developing an Offshoring Strategy

plan. This may involve determining the organization's *offshoreability* that could include an analysis of organizational, managerial, technical and cultural aspects along with an understanding of the human resources. Business drivers, competitive pressures and other market factors could also contribute significantly to the offshoreability of applications, systems and programs. Figure 2.1 depicting a waterfall model highlights some of the key steps that may have to be iteratively reviewed before building an offshore strategy. The activities described here extend from the need of businesses to move from an observer state to being a strong offshoring *player* as discussed previously.

Developing an offshoring strategy may involve a series of stages, depicted in Fig. 2.1, akin to the waterfall model of developing software applications. The key aspects of iteratively developing a sourcing strategy include:

1. **External Landscape Analysis:** Organizations typically begin their offshoreability analysis with a study of the external landscape which includes a review of the international business climate, country analysis and other global geopolitical and economic factors. Software service offshoring has been maturing in the past few years and globalization in IT is helping bridge the gap between developed and developing countries. (Ref: Appendix A—brief discussion on some of the popular sourcing strategies.) The primary driver behind offshoring is the availability of a pool of talented, and relatively inexpensive, IT workforce.

2. **Internal Analysis:** Organizational culture, business practices and operating environment are factors that could also define the shape of offshoring strategy. Organizations operating in highly regulated environments, like those doing work for governments on contract, or under other operational constraints may be a bit wary of offshoring their IT systems. Same may apply to companies with a strong culture of protectionism, trade unions or under other employment contracts where jobs may not be highly mobile. On the other hand, organizations already open to international business operations, those with joint ventures or subsidiaries overseas may be more receptive to offshoring. An analysis of the organizational culture, constraints and best practices will help build a business case for offshoring along with determining the scale of operations.

3. **Application Portfolio Analysis:** A portfolio analysis of the IT application stack should be undertaken with an eye towards identifying candidate applications that can be offshored with minimal disruption of operations. The inputs for this exercise can come from existing planning and documentation and the results could include rationalization opportunities and strategies for application portfolio consolidation. Some of the key steps in a portfolio analysis include

understanding the business functions and applications, scoring the application stack and determining the optimal offshoring strategy or tactics for offshoring. This exercise may unearth opportunities for application portfolio transformation and consolidation that may be synergized with offshoring initiatives.

4. **Offshoring Model Selection:** In the next section of this chapter we will examine the key offshoring models including Joint Ventures, Build-Operate-Transfer, Subsidiary models and sourcing to vendors. There are merits and pitfalls of each model and organizations may sometimes decide to adopt a hybrid model to suite their particular business need. Selecting the right offshoring model is one of the key steps in developing the sourcing strategy.

5. **Vendor Landscape Scanning:** Vendors have a key role to play in the success of an offshoring initiative. The selection of a supplier should be viewed strategically since many offshoring initiatives begin by attempting to leverage the cost arbitrage but move on towards a partnership model. A successful relationship can translate into enhanced productivity and quality, directly impacting the bottomline. Software service vendors, based on their maturity and experience may have experience in one or more offshoring models. During vendor selection, the organization may also have to consider the preferred model alongside the model that the vendor is comfortable with.

6. **Risk Analysis and Planning:** Organizations need to identify and evaluate offshoring and management risks early to avoid surprises and pitfalls. Risks in the offshore development models, could occur from various sources including internal organizational risks, technical risks and risks from the geopolitical landscape. Traditional project risk management techniques can be extended to address and mitigate some of the major risks identified during this phase.

7. **Piloting:** While it is important to articulate and strategize about the various facets of offshoring, it may be essential to pilot one or more models on a small scale before an organization and its people become comfortable with the workings. Piloting is essential for organizations embarking on offshoring for the first time and may also be done in phases with projects of incremental complexity being offshored. Piloting may also involve experimenting with different models and vendors.

Aspects of planning for offshoring are extended from general management, best practices of international business and strategy development, where the body of knowledge is vast and well developed. The best practices continue to emerge as companies begin to successfully operationalize offshoring on larger scales. The management of offshoring projects is fraught with unique challenges and risks and defining the offshoring strategy may depend on several factors including the organization's culture and even the geographic location of business.

OUTSOURCING: TWO SIDES OF THE COIN

Any strategy for offshoring will have to consider the existence of two distinct camps—that of the people and systems being sourced and the receiving end which will take ownership of the systems offshore. The two camps may either be in the same organization, in case of a subsidiary model, or may involve two distinct organizations—the offshoring company and a service delivery organization. The strategic goal of both camps may be the same, to ensure the success of the offshoring initiative, but the tactics and implications for teams and groups may be different. The teams at offshoring organizations may have to transition their work to offshore teams which can be a very unnerving prospect if the individuals transferring the work don't have another role or job lined up.

The management challenge in any offshoring initiative is to plan for a smooth handoff, knowledge transfer and transition of the offshored systems to a *steady state* with the least possible disruption. Managers at sourcing organizations need to plan to address the challenge of transitioning people who will be displaced or reassigned due to offshoring. The leaders at the receiving end need to plan to capture the processes, practices and know-how from onsite locations, and will have to plan for multicultural team building and development in addition to other project planning. The key goal of a smooth transition should be to keep it seamless and transparent to end users (of the IT systems) and other functional divisions and groups. Offshoring is as much about managing aspirations and goals as it is about outsourcing the development and maintenance of IT applications. Several factors come into play while planning a sourcing initiative and can include:

- **Managing People:** Project management and motivating people is a challenge even during the best of circumstances. Managing teams, individuals and their aspirations is one of the key challenges to be addressed while planning and executing offshoring projects. This could include aspects pertaining to displacement of jobs at the sourcing organization, and additions to teams at the vendor's end. People management begins during planning stage and continues after the sourcing initiative has begun moving towards a steady state. The inherent bonding of teams that occurs between groups adds a level of complexity when members have to collaborate with peers from across organizational, cultural and geographic boundaries. The success of any offshoring initiative is driven by the relationship, interaction and comfort of people at the client and vendor organizations. The expectations of people from both sides of the offshoring spectrum are going to be divergent and bridging the communication and intrapersonal gaps is perhaps the key to success.

- **Adoption of an Appropriate Sourcing Model:** The client organization needs to focus on adopting the best outsourcing model fitting their needs. Outsourcing vendors may either offer a portfolio of models or may have a preferred model they are comfortable with. Selection of an appropriate outsourcing vendor may hinge on the sourcing model that they have experience in delivering. Force fitting a model to a preferred vendor without prior offshoring experience should be avoided.

- **Strategy versus Operations:** Organizations need to formulate their outsourcing strategies while considering operating models, staffing management among other aspects of application management. Translating the business drivers (say, the need to lower cost) to operating models is one of the key challenges. Though the defining of sourcing strategies and operations may be done internally, organizations may benefit from unbiased advice from vendors, who because of their experience in the field, may be able to provide insights into the best practices and benchmarks.

- **Knowledge Management:** One of the key success factors of an offshoring initiative is the capture and transfer of operational, functional and technical knowledge. Though this sounds obvious, this is perhaps one of the harder aspects of sourcing since some of the processes, know-how, 'tips and tricks' and other aspects pertaining to the system being sourced may not be documented and may have to be gathered by observation. Knowledge gathering in an offshoring context may be harder when one considers the fact that the person or group that is likely to be displaced due to sourcing may not be as open to knowledge transfer. Offshoring vendors have addressed some of the issues of sourcing by developing their proprietary tools and techniques including the use of domain specific templates.

- **Program Management:** An operating framework that can guide and mentor ongoing projects, determine the project selection criteria and other aspects pertaining to the ongoing relationship between the client and delivery groups is an essential part of successful program management. Interestingly, this is an area of global delivery where conflicts are most likely to emerge since the stakeholders at both sides of the offshoring spectrum may have diverging goals. Managers at both ends need to work towards synergies that can satisfy their stakeholders. We will examine aspects pertaining to program management in further detail in the next chapter.

- **Abstraction of Communication:**   An important aspect of an offshoring initiative is an open, honest communication. Managers need to recognize the communication needs vertically across the projects and programs and horizontally to ensure transparency and continued management buy-in. This need permeates all aspects of the interactions between teams. While there is a need to create a transparency in all communication, levels of abstraction may creep in. The service delivery team may not be comfortable in exposing its internal workings and may focus on communicating the deliverables, work products and results. Similarly, the client may not wish to involve the vendor in all internal project planning and review activities. The risk of information overload is as great as the risk of lack of information. Both teams must recognize the need for a healthy 'yin-yang' and work within the confines of the operational model.

The basic offshoring model involves teams working onsite and offshore at client and vendor locations. The teams may have diverging goals during all the stages of a life cycle. During the planning stage, the onsite team may be concerned about the reduction in work, loss of familiar turf and the changes being wrought; the offshore team may be concerned about knowledge gathering, planning for a smooth takeover and other logistics. During the steady state, communication between teams will acquire significance and managers will have to ensure that the work being performed offshore is in accordance with the goals of the users of IT systems. A healthy yin-yang between teams and recognition of divergent goals will help members of both teams work successfully within the confines of the operational model. An awareness of some of the issues of sourcing and a deeper empathy for the goals of the other party, will help both the client and service delivery organization work towards a *win–win* relationship.

OFFSHORING MODELS

Offshore outsourcing of IT application development is generally a strategic decision rolled out in a phased manner. Organizations typically consider several inputs while formulating their offshoring roadmap, and selecting an appropriate offshoring model is perhaps one of the key steps in the process. Selecting the right model can translate to success of the offshore outsourcing strategy and ensure a buy-in from all the stakeholders. While formulating and choosing an offshoring model, there are several factors to consider, including aspects of international business strategy—selecting the country, scanning the landscape and deciding on the sourcing strategy. Three models most popular among business leaders and strategists generally include:

1. Joint Ventures
2. Subsidiaries/Captive Development Center
3. Sourcing to Vendors

A brief discussion of each of these models follows.

Joint Ventures

A Joint Venture (JV) is perhaps the most popular model of offshore outsourcing whereby an organization ties up with a local firm or company either by taking an equity stake or forming an independent company by jointly contributing resources. The goal is generally to work towards a 'win–win' deal where both organizations hope to benefit from the other's strength. The foreign company brings in its processes, best practices and technical know-how and the local firm provides its expertise of the domestic market, knowledge of local practices and hiring talent. Some of the highlights and benefits of a Joint Venture include:

- *A Well Defined Business Plan:* Successful JV agreements begin with a sound business plan and documentation of the expectations that includes establishing mutually beneficial terms of operation and goals. Both parties should set realistic and achievable expectations and protocols of communication. The goals of both parties should be acknowledged in the business plan. The outsourcing organization aims to gain from lowering total cost and other benefits of sourcing. In return, the vendor organization looks towards international exposure and an opportunity to learn some of the best practices and operational aspects of working with other global organizations.
- *Legally Binding Contracts:* JV are typically incorporated as a legal entity or operate under sound agreements and contracts that help in maintaining continuity of business and operations. The terms of the contract, legal jurisdiction and other aspects of performance including the Service Level Agreement may also form a part of the contract. A contract with an offshore entity can be nebulous and may need inputs from legal experts in multiple countries and provide for mechanisms of dispute resolution, arbitration and settlement in case of disagreements.
- *Retention of Control:* Unlike other sourcing models, the client or sourcing organization in a JV can retain considerable control over the outsourcing initiative since it will also own a direct stake in the enterprise. Owning a stake ensures

that decisions from the top can be driven by the client organizations. This also stems from the fact that the joint venture will only serve the interests of the client and not of any competitor.

- **'Win–Win':** In a JV, both the customer and the vendor organization bring something to the table and both bring opportunities. The client brings the background of managing the processes, industry 'best practices', technical know how and other managerial expertise, whereas the vendor brings in the talent and local pool of expertise along with a knowledge of cultural and operational issues pertaining to the local market. By successfully managing a JV, both parties have an opportunity to profit and learn from each other.
- **Risk Mitigation:** The biggest advantage of a joint venture is the opportunity for both parties to mitigate the risks by focusing on their core strengths while relying on the other party to bring in expertise in areas where they are not strong. The sourcing organization attempts to transfer the operational risks to the domestic partner while gaining knowledge of local markets; the vendor organization also gains by being shielded from international business and marketing risks.
- **Operational Issues:** Although there are several benefits of a JV, both parties in the deal need to tread carefully. As there are diverging interests, the client and vendor organizations need to establish an environment of trust and continue to monitor and manage the operations to work towards mutually beneficial goals. Depending on the circumstance, one of the parties may end up becoming a 'minority' stakeholder, in which case it should make a conscious effort to continue to maintain an element of control at the managerial level.

A JV model is a relatively low-risk strategy adopted by organizations entering offshoring in a new country and market. By capitalizing on the strengths of a local player, the client organization can mitigate some of the risks of internalization; similarly, the local player can benefit from partnering with a strong player and the opportunity to scale up the value chain.

A JV contract may sometimes include build, operate and transfer clauses to motivate both parties to work towards a clearly defined exit strategy. Build-Operate-Transfer[1] (BOT) and Build-Own-Operate-Transfer[1] (BOOT) may involve an option for the domestic company to sell its stake to the foreign company after a stipulated period or after agreed upon milestones are reached.

Subsidiaries/Captive Development Center

The operation of a successful JV may involve constant relationship management and adherence to the diverging interests of both parties. Also, the joint venture may sometimes have a Build-Operate-Transfer clause in which case, the operation may transfer its ownership to the vendor organization after a specified amount of time or after certain milestones have been attained. Organizations may sometimes decide to bypass the JV model altogether and directly go in for a subsidiary or local-office if the management is comfortable in dealing with the nitty-gritty of internationalization and local market operations. Some of the popular terms used to describe the model include Offshore Development Centre (ODC), Captive Development Center or in some cases simply *branch* or *local office*. The popular modes of establishing subsidiaries in the IT service management context include setting up local offices, subsidiaries or alliances and 'virtual' JV. From an operational management standpoint, subsidiaries operate as independent business units or branches, executing programs and projects for onsite teams; from this perspective, the mode of managing a subsidiary is similar to managing projects and programs in a Global Delivery Model (GDM) promoted by software service delivery organizations.

The local office model is extremely popular among hi-tech organizations that are comfortable in management of technology development and innovation and look to offshoring as an extension of their diversification strategies. Most of these organizations already operate their business units in a decentralized mode with individual lines of businesses operating in different sites even in the home country. Therefore, moving to an offshore location merely extends their diversification strategy while capitalizing on the benefits of operating in low-cost geographies. This model is also popular among businesses that already have global operations. Migrating IT application development to local offices, where it makes the most economic sense, is an extension of their globalization strategies.

Large software development companies including IBM, Microsoft and Oracle are already comfortable doing business in a global marketplace; for them moving development or maintenance of some of the projects and work is a way to extend their geographic footprint. Similarly, software giants like Accenture, EDS and Braxton, among others, have been at the forefront of bundling newer services for their clients, offshoring being the latest in their suite of services. Many of them already operate in multiple geographies and for them to operate local offices providing captive development to clients is a natural step in the evolution. Interestingly, some of the large offshoring vendors also use the term Offshore Development Center—a model we will talk about in the next section—interchangeably with the term local office to imply a subsidiary for clients.

The key challenge in a subsidiary model, apart from internationalization and localization of business management, includes aspects pertaining to management of expatriate staff, line workers, technical experts and line managers from multi-cultural backgrounds. Issues such as cross cultural management, communication gaps and head-office vs branch-office syndrome can surface easily and planners need to guard against them upfront. Other issues of internationalization such as adherence to local laws, customs and mores, business and work practice also need to be closely planned while formulating the operational strategies for local businesses.

Year	Entry of EDS into Indian market
2001	EDS–India opens its fourth state-of-the-art facility in Chennai at Tidel Park.
2000	EDS–India opens new office at DLF Plaza Tower, Gurgaon. India Solution centre achieves CMM Level 3 in July.
1999	EDS–India awarded the ISO 9001 Certification by KPMG in February.
1998	First center opened in Chennai, the commercial capital of southern India. Communication network linking EDS-India to EDS worldwide becomes operational in May. Curtains up on the second state-of-the-art center in Chennai in July. Quality initiatives for ISO 9001 roll out in August.
1997	Strategy Development. Finance committee approves EDS-India's three-year plan.
1996	EDS–India shapes up as a 100 percent-owned subsidiary of EDS. EDS–India becomes first company in India to sign a multiyear outsourcing contract.
1995	EDS–strikes relationship with vendors in India. EDS–sets up liaison office in New Delhi.

Table 2.1   Sample outsourcing model for entry into the Indian market

(Source: EDS.Com)

There are several strategies available to mitigate the new market risks including a phased entry, forming joint ventures with vetted local partners. An example is the multi-stage entry of EDS[2] into the Indian market. (Ref: Table 2.1).

Sourcing to Vendors

The joint venture and subsidiary models of sourcing that we looked at in the previous sections may involve deep commitment on the part of a sourcing organization, a move that managements may sometimes be averse to. To counter the perceived risks of these models and to capitalize on the benefits of offshoring, companies resort to outsourcing projects, programs and individual work orders to offshore vendors in a low-risk entry strategy. Interestingly, sourcing to vendors is also the most visible offshore outsourcing model and encompasses a wide range of work ranging from sourcing small projects to multi year contracts amounting to millions of dollars.

In typical offshore outsourcing, an organization sources a defined work item to one or more pre-qualified vendors specializing in those technologies. The vendor generally takes on the management of the development and delivery life cycle and the clients dedicate some resources and people to *'program manage'* the work. Some of the popular forms of outsourcing to offshore vendors include:

1. ***Onsite subcontract (with offshoring):*** Most offshore outsourcing firms trace their history back to their software services mode and continue to offer onsite project support along with some staff supplementation. Onsite subcontract with some offshoring is perhaps the least risky outsourcing option available to the client since it mitigates communication and other risks; however, it is also the most expensive outsourcing model.
2. ***Pure Offshore Projects:*** This model of offshoring is less prevalent and generally seen only in a small scale development of software component or modules. Although we document this model for completeness, 'pure offshoring' of projects is more of a misnomer since it is a model that companies or offshoring vendors have not really perfected (yet). Software requirement gathering, designing and

translating such design into workable code involves a level of abstraction and fuzziness even at the best of times. To expect software teams to gather requirements and hand them over to some other team halfway across the globe and expecting the offshore team to work in isolation is a level of abstraction that is yet to take off. Pure offshoring of projects is a model generally combined with other models like Onsite Subcontract.

3. ***Offshoring Individual Projects:*** Managers at client organizations who have well defined deliverables, programs or modules to be developed source them to vendors. In some cases, they don't go through a formal bidding and vendor evaluation process; instead, they post their needs on online web portals inviting programmers from around the globe to bid for their work. A case in point is the 2RentACoder portal[3] (Ref: Box 2.1) Although this model has really not taken off in a big way in the corporate world, a more connected global economy is a precursor to this trend catching on.

4. ***Global Delivery (Onsite/Offshore) Model:*** This is the classic offshoring propagated by most software service vendors. In this model, the vendor takes on the project, module or program from a client organization, deploys a small team onsite that works with the client managers and teams and co-ordinates work with the offshore team that does the bulk of the work. Operationalizing the workings of the onsite and offshore teams is perhaps the key success factor of this model. This is also depicted in the sweet-spot of optimizing on the risk and cost curves (Ref: Fig. 2.2) involving managing an optimal staffing model during the project execution.

5. ***Offshore Development Center (ODC):*** An offshore development center or ODC is the backbone of global delivery and is sometimes known as a captive development center. Software service companies are increasingly setting up captive centers in low cost offshore destinations including

Ireland, India, the Philippines and elsewhere. They also employ smaller teams of people onsite at their nodal offices and at *proximity development centers*. Interestingly, the term ODC is used by both client and offshore service organizations to refer to their branch offices. Connected to the main head-office or nodal offices by means of secure, high-speed networks, these ODCs provide advantages of diversifying operations while providing for some autonomy in operating branches.

6. *Multi-vendor Offshoring:* In the discussion on offshoring models that we looked at thus far, we assumed the relationship between a client and a vendor. However, in reality, a client may have multiple vendors working on a project or initiative. Organizations attempt to de-risk their outsourcing strategies by empanelling a selected list of vendors (a.k.a. preferred vendors) from which individual projects and managers opt to select and source work. This strategy allows enterprises to consolidate their broad sourcing strategies while ensuring consistency across projects. This also gives individual managers some flexibility to select a particular vendor or vendors suited for their custom needs. This also breeds some competition among vendors while reducing dependency on a particular vendor. A case in point is GE's famous 70:70:70 rule that was originally articulated by its former Chairman, Mr Jack Welch. The vision statement said that 70% of a business unit's processes should be outsourced, 70% of the outsourced processes should be offshored and 70% of the offshore outsourced work should be done in India. (Ref: Chapter 3, Box 3.2).

Offshoring can be a complex and strategic decision. Since it is hard to change course midstream, organizations and business leaders need to spend considerable time strategizing and planning the model suitable for their specific needs. The models highlighted in the discussion are some of the most common ones encountered and may not fit every business scenario or need. Clients in the West are learning about the pros and cons of the different models offered by players in the marketplace, in some cases specifying a hybrid model tailored to their businesses. This is especially true of large enterprises that may adopt multi-vendor strategies with one or more models tailored to specific business needs or to address the certain geographic operations. Many large outsourcing vendors also provide a mix-and-match portfolio of options to their clients and sometimes draw a roadmap to migrate from one model to another as the client's understanding of the offshoring business matures.

Box 2.1

MICRO-OFFSHORING: INTERVIEW WITH THE CEO OF RENTACODER[3].COM

In this section, we present an interview with Ian Ippolito, CEO of 2RentACoder.com to examine some of the aspects of niche offshoring.

Question: Q, Ian: I

Q: What is the impact of the 2RentACoder model on Project Management of globalized teams? Do you see managers using this as an extension of their existing teams to supplement talent and work?
I: *Yes managers do use RAC to extend their teams and farm out extra work they don't have time to do.*

Q: What are the key project management skills involved in managing projects using freelancers?
I: *Here are the key skills I see being crucial to RentACoder for buyers managing projects:*
    *1. **Skill in dealing with cultural differences:** Successful buyers understand how their culture differs from their coders and take steps to avoid problems arising from that.*

Box 2.1

Micro-Offshoring: Continued...

As an example, most Indian coders are very customer service oriented and very respectful to the buyers. The flip side of that is that many often don't like to tell the buyer 'no' even if they know a particular idea from the buyer doesn't have a high probability of working out. This also extends to not liking to give the buyer 'bad news' in their status updates if the project isn't going well for them. This can cause a big surprise problem for the buyer, come deadline time, if the coder isn't able to 'fix' things before then.

If the buyer is a US buyer, their first reaction may be to get very upset. In the US employees are more straightforward/ blunt about communicating bad news to their bosses. An American buyer who was unaware of the difference, might think that the Indian coder was lying or withholding information or was incompetent. However the American buyer who understands cultural differences probes more carefully and deeply when he hears a positive status report or a 'yes boss, that's a good way to go' to make sure that there isn't something potentially wrong that is not being addressed.

2. Attention to detail/follow-up: An outsource project involves less regular communication than an insourced one and, as a result, has a higher potential for problems. It's crucial that a buyer manage the project by paying attention to all details. Daily follow-ups are crucial and each milestone or mini milestone must be monitored closely.

Follow up skills are crucial when a key milestone is missed. Some new buyers make the mistake of not tightening down control when this happens, thinking that the coder will 'do better next time'. This is usually a big mistake, and usually results in another milestone being missed. The buyers who have success with a missed milestone are usually the ones who take proactive steps to force the coder to increase

Box 2.1

MICRO-OFFSHORING: CONTINUED...

their participation, communication and commitment to the project. They do this by increasing the required frequency of communication and creating more discrete and detailed milestones.

3. Flexibility: The coder is often in a different time zone than the buyer. This can make communication more challenging. Successful communication usually requires the buyer accommodating a less convenient time when obtaining status reports, to co-ordinate with the freelancer's timezone.

4. Willing to invest large chunks of time: Some buyers (especially new ones) don't realize that custom software will only be as good as the amount of time they invest in communicating with their coder. It's like building a custom house. If someone says 'just build me a house' the house they end up getting is almost 99.99% certain to NOT be even close to the house they 'imagined' they'd be getting. The smart house buyer sits down with the builder and goes over the blueprints to make sure each room is right, and then each wall, and then the fixtures, and so on. That way the end result is exactly what they wanted. It's the same thing with managing a software developer.

Communication is a key factor in managing globalized teams. Projects periodically fail because of poor communication, especially in a cross-cultural context. However, it is difficult to generalize about a strategy that applies to all countries, other than that I recommend buyers learn as much as possible about the culture of the coder(s) that they are working with. The Culture Shock series is an excellent set of books that really allow buyers to understand the differences, anticipate problems and work around them.

BOX 2.1

MICRO-OFFSHORING: CONTINUED...

Q: Another emerging trend/paradigm in software develop-ment is the open-source wave. Do you see that as a threat or an opportunity?

I: *Open Source is not a threat to Rent a Coder as it might be to a company like Microsoft that makes proprietary products. It is actually an opportunity for us. We actually do quite a bit of business with people wanting to customize open-source products for their own uses, products such as OSCommerce for example.*

The flip-side of open-source (for Rent a Coder anyway) is that there sometimes isn't a good understanding of the copyright ramifications of its use by coders (especially, and understandably, those from countries where there aren't strict, or any, copyright laws). As a result, sometimes coders get into trouble by building off of open-source products and trying to sell them to buyers. That doesn't fly.

Q: Are there any other questions/answers you wish to address on the theme of 'globalization and economic environment'.

I: *Right now there is a backlash against globalization occurring here, particularly the offshoring of white collar jobs such as computer programming. The perception is that this harms the US economy. However there are two other sides to it that people don't see on the news. One is the enrichment and thanks that some of these coders have for the opportunity to work. A coder in Romania wrote me a few months ago and said he actually had saved up enough money over 2 years of working on RAC to buy a car, something not many Romanians can afford. The second, is that here in the US I've seen first hand the birth of an uncountable number of small US companies that never would have even had a chance to get off the ground without access to the global marketplace for programming. They simply couldn't afford to be in business if they had to pay the higher rates that used to be norm.*

RISKS OF OFFSHORING

Managing work being done halfway around the globe is not without its risks or challenges. Enterprises entering into offshore development relationships with vendors or establishing subsidiaries will require the ability to respond quickly to changing business needs, diverse cultural differences and software development infrastructure risks including network security issues. Articulating the risks of managing Offshore IT Development Projects, Ralph Kliem[4] says *"Offshore outsourcing of IT development projects does not eliminate risks. Instead, it combines new risks with existing ones often associated with onshore projects. Project managers must recognize that these offshore projects will likely require even greater management of risk due to the unique challenges posed by geographical, cultural, and other differences. Otherwise, the gains attributed to offshore outsourcing of these projects will not be realized, and the chances for failure may be augmented."*

Decision makers need to balance the basic offshore business benefit of low cost versus project and operational risks. With offshore development, risk management assumes a renewed focus. Risk in the offshore development model is contributed by three major sources:

1. Organizational Risk
2. Technical Risk
3. External Risks

Organizational Risk: Global Business Risks

IT projects are fraught with risks, and risk management and planning is a significant aspect of project management. In addition to the planned project and program risks, unforeseen organizational and business events may also trigger risks during execution. The study of international business management focuses on risk management, especially on aspects distinct to globalization. Aside from the risks of global business management, there are a number of risks inherent to offshore outsourcing, including:

- *Cultural and Distance Issues:* Perhaps the biggest challenge of managing offshore development is the need to manage teams from across diverse geographic, cultural and ethnic backgrounds. Working with teams spread across such geographic and national boundaries could potentially increase the risk of misunderstanding and result in project delays and unnecessary rework. Managers of global delivery projects need to focus attention on such risks arising out of culturally and geographically disparate teams.
- *New Market Risks:* Operating businesses in new markets and geographies have certain inherent risks attributable to local and international business practices including legal, regulatory and other environmental issues. Companies entering into outsourcing relationships overseas will have to weigh the risks and rewards before committing to any long-term investment. There are several international business management strategies that address and mitigate the new market risks including a phased entry, forming joint ventures with vetted local partners or working with contractors. Offshoring software systems can reuse some of the best practices from international business management to address the new market risks.
- *Perceived Loss of Control:* Sourcing development of applications to distant locations may lead to some loss of physical and hands-on visibility. Such loss of visibility may sometimes be misunderstood as a loss of control, especially among programmers and software specialists who are known for their *'geeky'* outlook towards teamwork. An initial perception of offshoring among team members may be that of ceding of project control. To mitigate this perceived risk of loss of control, the Management needs to

clarify goals and expectations at the outset and simplify the channels of communication.

- **Compromising Confidentiality:** Protecting intellectual property, internal processes and methodologies is among the prime concerns when companies expand operations overseas. Turnover of employees, intellectual property loss and disaster recovery are real risks inherent in any international business venture. Added to such concerns are disparate laws on intellectual property, patents and other copyright protection. To address such concerns and encourage globalization, countries are beginning to enforce intellectual property laws and the WTO's mandates stringently. Managers need to plan for and adopt various measures to protect their proprietary processes while moving offshore. This may include physical, network and other security measures, enforcement of laws and other means to protect IP and confidentiality.

Technical Risks: Challenges for IT Staff

The challenge facing the development staff in offshore development projects is significantly greater than domestic in-country development. Steve McConnell[5] says, *"The stereotypical programmer is a shy young man who works in a darkened room, intensely concentrating on magical incantations that make the computer do his bidding…. Vital information is stored in his head and his head alone. He is secure in his job, knowing that, valuable as he is, precious few people compete for his job."* This essentially translates to the challenge of managing knowledge of systems that is traditionally done by enforcing system documentation, user manuals and the like. It is for the managers and the offshoring team to address the aspirations of technical people and address issues pertaining to offshoring including the following:

- **Onsite/Offshore Communication and Co-ordination:** Communication between and across teams and projects is perhaps the biggest challenge managers face. Co-ordination of onsite and offshore development activities may require additional focus and enterprises must dedicate increased management resources to ensure seamless communication. Simple techniques of communication like brainstorming and meetings can become harder due to time-zone differences. Getting people from different time zones together in a *virtual* meeting session can be a real challenge for managers. A framework for communication, similar to the communication layer prescribed in the proposed Offshoring Management Framework (OMF, Ref: Chapter 3) will help mitigate the risks of co-ordination between offshore and onsite teams.

- *Limitation of Management Tools:* Most of the tools and techniques of software development like version control, issue tracking and management are designed for software development in a single location. Managing global projects requires teams to begin exploring newer tools and techniques that can mitigate the associated cost of administration and management of offshore outsourcing projects.

- *Infrastructure Issues:* Successful application development projects need reliable infrastructure, hardware and networks. Increased complexity of network security, network operational issues and disaster recovery/backup requirements are further exacerbated by offshoring. Sophisticated tools and technologies along with 'best practices' including tools of collaboration and communication help mitigate some of the infrastructural challenges of offshoring.

- *Knowledge Management:* Managing project information, data and communication repositories is one of the key success factors behind managing application delivery. This includes managing the data flow and repository of information including project history, best practices and use of tools and techniques. Offshore development implies a greater reliance on Internet based technologies as the framework for software development infrastructure. The bulk of written communication will be in the form of emails and transferred files. Use of knowledge management tools and techniques can help mitigate the risk of both data loss and information overload.

External Risks: Global and Geopolitical Risks

Offshoring, like any international business venture has to address and factor external and geopolitical risks. The outbreak of the SARS virus in 2003 (Ref: Box 2.2) brought home the implications of doing business in a globally connected world. These risks include:

- *Geopolitical Risks:* When IT managers take business decisions, they review the technology, services, personnel and company risk. While managing global projects, they also have to consider political, geographic and geopolitical risk as well.
- *Regulatory and Governmental Restrictions:* Like geopolitical risk stated above, offshore outsourcing can sometimes be hindered or aided by governmental policies and laws. What makes this risk more challenging is that laws of both the outsourcing and outsourced organizations, along with other international regulations come into play. This includes restrictions on travel, immigration and visas that constantly haunt players in the outsourcing industry.
- *Currency, Global Business Risks:* Volatile exchange rates can have a significant impact on the profitability of any international project. This is also true of offshoring ventures.

The risks highlighted in this section are by no means exhaustive. Individual organizations, business verticals and geographies may have their unique risks that project planners may have to address. Alongside the examination of risks, the expectations, deliverables

Box 2.2

CASE IN POINT: SARS AND THE MANAGEMENT OF GLOBAL RISKS

Two distinct events in early 2003, the outbreak of the SARS virus in Asia and the declaration of the second Gulf War on Iraq, directly impacted international travel, in turn posing a major threat to offshoring. Business leaders across the globe intuitively cut-back on global travel, especially to the Asia Pacific region while business magazines and newspapers began speculating the short-term and long term impact of trends in articles provocatively titled, "SARS seen hitting IT sector hard."

Although the outbreak of SARS or the war in Iraq had little to do with IT development or offshoring, it had the potential to scuttle businesses and projects since it impacted international travel, which continues to be the lifeblood of offshoring. IT companies and service delivery firms began to step up their contingency plans to ensure business continuity with minimal travel. Though the companies gave financial and earnings warnings, many managed to come through with minimal impact on projects. For instance the India offshoring firm, Wipro[6] in a case study reports *"Although, Hong Kong and Singapore were affected by SARS (Severe Acute Respiratory Syndrome) in March and April 2003, there was no delay in the launch of the project as various measures were taken to counter the effects of the calamity."*

Even with modern communications technologies including emails, instant messengers, video conferencing and teleconferencing, the need for in-person meetings, client visits and personal business negotiations continue to involve people traveling across the globe. Over a year after the outbreak, management thinkers continue to ponder over the implications of other similar outbreaks and the unforeseen consequences thereof. Robin Mckie[7], in a recent article, says, *"A flash of a boarding pass and the virus, already multiplying in the cells of human hosts, moved on to new, infectious pastures."* The SARS incident brought home the realization among business leaders and technologists that offshoring initiatives are fraught with unforeseen geopolitical risks that they need to plan for and address.

and service levels of offshoring have to be set upfront to minimize any surprises. There may also be other hidden costs, control and co-ordination issues that may contribute to management risks that the offshoring governance and planning team needs to consider. Use of appropriate tools, techniques, best practices and technologies will help organizations and managers address and mitigate any risk accruing from outsourcing. Without a clear solution to some of the challenges and risks articulated above, managing offshore software development projects may prove difficult.

SELECTING THE OFFSHORING MODEL

Popular offshoring models include Joint Ventures, subsidiaries, sourcing to vendors and include strategies of hybrid models that organizations adopt after examining their business drivers and risks. The process of selecting an appropriate offshoring model can be a challenging process. Researching foreign markets, studying the costs and benefits of offshoring and deciding on the right model can be daunting to many businesses. Sensing an opportunity to extend their suite of services, most large offshoring vendors and IT service organizations have begun to dedicate teams and researchers to study the successes of models adopted by their clients and actively benchmark them. Many vendors share such prescriptive models from the field with their prospective clients. Research firms including the Aberdeen Group, Gartner, Giga, McKinsey's Global Institute and others have also dedicated considerable bandwidth to researching the offshoring models and facilitating clients. In the previous section, we looked at some of the common offshore models prevalent in the industry. However, we only examined the models from a high-level operational standpoint. Some of the inputs for selecting an offshoring model include:

- **Offshore/Onsite Mix:** Selecting an appropriate model will depend on the offshorability of an initiative, essentially the mix of onsite and offshore personnel most appropriate for the execution. One of the key aspects of selecting appropriate offshore outsourcing models is the examination of cost and risk in the offshore/onsite mix of resources in a project. As the number of resources onsite increases, the cost goes up; on the flip side, as the number of offshore resources increase, the risks of management and communication increase. Most companies look for the optimum models, basically the point of intersection of the risk and cost curve (Fig. 2.2). Most vendors claim to use their proprietary offshore or Global Delivery models that help them optimize the team sizes, sometimes to the tune of 80:20 or 70:30 (Offshore:Onsite).
- **Risk Tolerance:** Not all businesses are equally tolerant of risks. An analysis of the organizational, technical and external risks may be done to define and examine the relative ranking of the different categories of risk. Risk need not only be with respect to operations or stability, it can also be the perceived risk of security, control of management and other aspects pertaining to business sustenance.
- **Scale of Operations:** The overheads in starting and operationalizing a subsidiary or even a joint venture may sometimes drive enterprises, especially smaller organizations

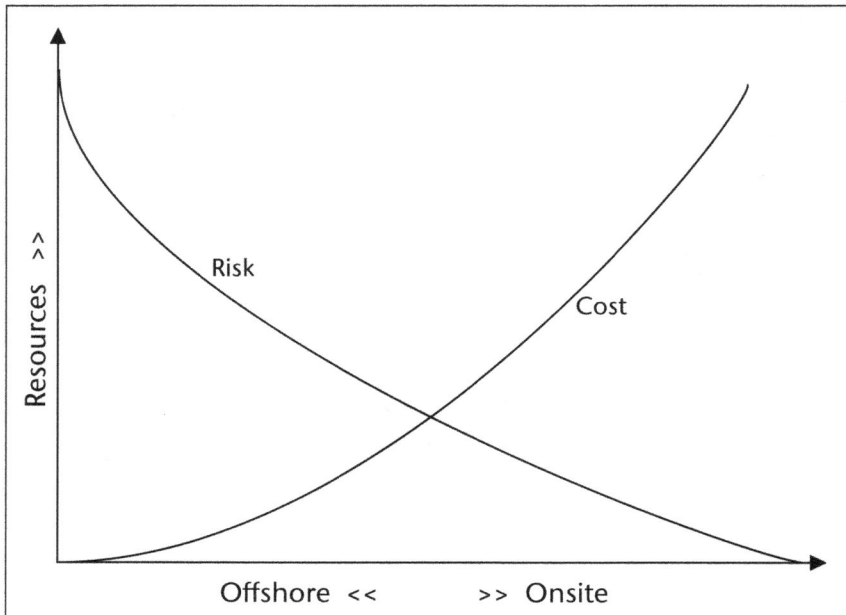

Fig. 2.2   Typical Cost-Risk balancing in Offshore/Onsite selection

with fewer projects to offshore, towards direct sourcing to vendors. On the other hand, large companies that hope to transfer a sizable number of projects offshore may benefit from the economies of scale offered by operating a subsidiary or a BOT model.

- **Comfort in International Business Management:** Many organizations, even large companies are not very comfortable operating across geographic and cultural boundaries. For them, operating a foreign subsidiary or joint venture halfway across the globe will sound more ominous than sourcing to a vendor. On the other hand, organizations that are already comfortable in operating in multiple geographies may prefer to manage their own captive offshore development centers.
- **Strength in Vendor Management:** The different offshoring models require varying degree of working with

vendors and subsidiaries. Outsourcing work, by nature, requires the sourcing organization to develop and manage a strong vendor-management process.

The risk and cost curve highlighted in the figure above focuses primarily on the risks associated with offshoring. It may be argued that other organizational or project management risks may mitigate the risks of offshoring and may become primary drivers in selecting an offshoring model. The factors highlighted in the discussion above are merely indicative and the actual decision making process will be much more complex. It may involve a detailed analysis of factors of international business management, country analysis, cost-benefit analysis, study of legal and other operational factors, vendor analysis and an internal management study. The section on Global Delivery Framework later in this book delves deeper into aspects pertaining to the operation of global delivery that will also be a factor to consider in deciding on the offshoring model.

NOTES

1. The Build-Own-Operate-Transfer, BOOT, model of outsourcing is popular among governments and organizations with large infra-structural projects. This is gaining prominence among IT companies wishing to mitigate risks of offshoring. [*The Hindu Business Line*, May 28, 2003]

2. EDS [http://www.eds.com/]

3. 2RentACoder.com is an online exchange that caters to the custom software programming requests from users across the globe. Buyers, individuals or corporations, who need some software or programs developed, pick from a pool of over fifty thousand coders. Coders view projects from across the globe and place their bids on what they would charge to do the work. The buyer can browse through each bidder's resume, and when they find a coder they consider fit, they can hire them instantly.

4. Managing the Risks Of Offshore IT Development Projects [Ralph Kliem; *Information Systems Journal*, Volume 21, Issue 3.]

5. Section: Orphans Preferred [Steve McConnell, *Professional Software Development*, Aw Professional]

6. Wipro [http://www.wipro.com]

7. "Nature, the most deadly bio-terrorist of all" [Robin Mckie, *New Statesman*, 1 January 2005

CHAPTER 3

Framework for Managing Global IT Projects

- Offshoring Management Framework (OMF)
- Governance Layer
- Service Level Agreement (SLA)
- Transitioning Offshoring
- Managing Offshoring Programs
- Conclusion

Offshoring models and strategies in the IT industry are evolving and are beginning to show signs of maturity. While client organizations are beginning to consider offshoring as a strategic tool for management of IT applications, Software Service providers are also getting comfortable taking on larger, geographically distributed projects. They are positioning their proprietary development models as differentiators in the marketplace. '*Global Delivery Model,*' '*development follows the sun,*' '*24 X 7 delivery,*' '*Strategic Outsourcing,*' and '*Offshore Outsourcing Model*' are some common phrases used in the nascent industry. Some of these delivery models from vendors are being positioned to be proprietary, and there is a common thread running through the building blocks. A case in point is a query from a client team on an 'offshoring study mission' to India[1]:

We would like to leave with a better understanding of the delivery models of offshoring and would like to 'see' it in action. Specifically we want to understand:

1. *The communication model and management of an offshore project.*
2. *What changes we need to make in our process and project execution models to work with an offshore provider.*

This brief perhaps sums up the essence of where many outsourcing organizations are coming from. Enterprises with existing IT infrastructure, management systems and processes in place are looking to leverage the offshore project execution capabilities of service providers and vendors and in some cases set up their own subsidiaries. Executives are also realizing that executing and delivering IT projects involve *regular* challenges that they already have processes to address; global delivery is an added dimension to the management imperative. This new dimension itself involves several unknowns and challenges including managing efforts and schedules of individuals, teams and resources from across the globe. This also includes the need to acquire knowledge of practices to ensure smooth transfer of work and work-products between offshore and onsite teams.

While researching the emerging trends in globalization of project management and offshoring, it was obvious that although several service delivery organizations use proprietary processes to manage offshoring projects spanning geographic and cultural boundaries, the core of their processes follow distinct patterns. The models adopted by service delivery vendors and other offshoring organizations derive from the published body of knowledge of project management and the application development life cycle. Best practices of globalization, internationalization and managing the workflow of teams and group dynamics also extend into such practices. We will attempt to look at some of the intricacies of offshoring taking a vendor neutral stance with a specific emphasis on managing and executing projects and initiatives using what we shall call an Offshoring Management Framework (OMF).

The underlying assumption behind the OMF is that the generally understood Body of Knowledge, industry best practices and organizational processes will continue to form a basis for most of the IT managerial and planning and developmental functions. The Framework borrows extensively from other publicly available sources including published articles, whitepapers and corporate websites. Although we draw on experiences in managing global delivery projects, we shall consciously attempt to exclude any references to an individual organization's proprietary know-how and practices. References to Infosys' (my employer) business and operating models will be restricted to publicly available pre-published material.

OFFSHORING MANAGEMENT FRAMEWORK (OMF)

Large Businesses may follow several paths towards offshoring sometimes using a top-down or a bottom-up approach towards sourcing. Some organizations follow a bottom-up approach by allowing individual IT division and groups to try outsourcing at individual project and program levels before formulating an organizational strategy. In such an approach, the organization may have a few internal offshoring experts, essentially managers who learn the intricacies of sourcing by sending a few projects offshore. Over a period of time, such internal experts may be called in to provide inputs to other project teams and groups. However, offshoring experts in the bottom-up approach may not have formal authority to set standards for use by others but act more as a point of reference. A few organizations like General Electric[2] articulate their corporate sourcing strategy formally; such strategies may be bundled with their overall IT planning and strategies which they roll out to the different Line of Business units. Managers from the planning group may form a part of a Program Management office and may have formal authority to define and articulate an org-wide strategy.

Box 3.1

CASE IN POINT: CISCO'S OFFSHORING THRUST

Cisco, an American Telecommunication giant with a market capitalization of over 127 billion dollars, has been at the forefront of offshoring. As a premier provider of telecommunication products and applications, the company's core product line is the backbone of many outsourcing ventures. Outsourcing organizations, vendors and service firms depend on Cisco's routers, telecom switches, Voice over Internet [VOIP] and other technologies to enable seamless communication.

The company has also been at the forefront of outsourcing the development and maintenance of its IT and business processes to offshore locations. Mr. S. Devarajan, Vice-President of Cisco Systems India, was quoted[3] saying that during the year 2002, Cisco offshored more work to its three Indian partners, Infosys, Wipro and HCL Technologies while also sourcing software work to vendors like Zensar Technologies. The company is also outsourcing IT-enabled services to the Infosys subsidiary, Progeon. Besides sourcing work to vendors and partners, Cisco has also set up its own Offshore Development Centre in Bangalore, India, primarily to outsource product research and development. The company also has globalized development and research by establishing R&D Center in Japan.

The company that was known for its world class acquisition strategy during the dot.com era is probably not using the offshoring model of either of its offshoring partners or vendors in isolation. The company has not made its offshoring strategy or details of the models it adopts public and there are not many published case studies on its sourcing strategies. However, one can infer that a complex web of managing multiple vendors and its own offshoring would require a unique strategy.

In many organizations, offshoring is a reaction to changing business dynamics and not a formal strategic initiative. Even in such instances, organizations and managers don't have access to formally published sourcing models and generally depend on their vendors to proactively suggest approaches. As offshoring strategies may survive beyond the life of relationship with a vendor, managers and business leaders need to take a vendor-neutral approach towards sourcing and eventually work with vendors who are comfortable with other popular models.

The proposed Offshoring Management Framework will attempt to address some of the most common managerial issues and challenges faced by companies that are evaluating models offered by different vendors while formulating their individual strategies. It may be used as a frame of reference for planning projects spanning geographic and cultural boundaries. The management imperatives in global delivery of software applications include buy-in and sponsorship from senior executives, program and project management, managing the operation of development and maintenance life cycle and creating an environment of open communication across the teams. The OMF, depicted in Fig. 3.1, will attempt to address four major areas of focus:

1. Governance Layer
2. Management Layer
3. Project Execution Layer
4. Communication Layer

Typical executive challenges include defining a strategy and managing the transition and the steady state. The Governance Layer addresses these challenges faced by executives of both the offshoring and the vendor organizations. In the governance layer we also address aspects pertaining to SLAs, managing the transition offshore and Program Management. Projects are typically a self contained work effort targeted at solving specific problems and, in an offshoring context, the smallest unit of work that may be

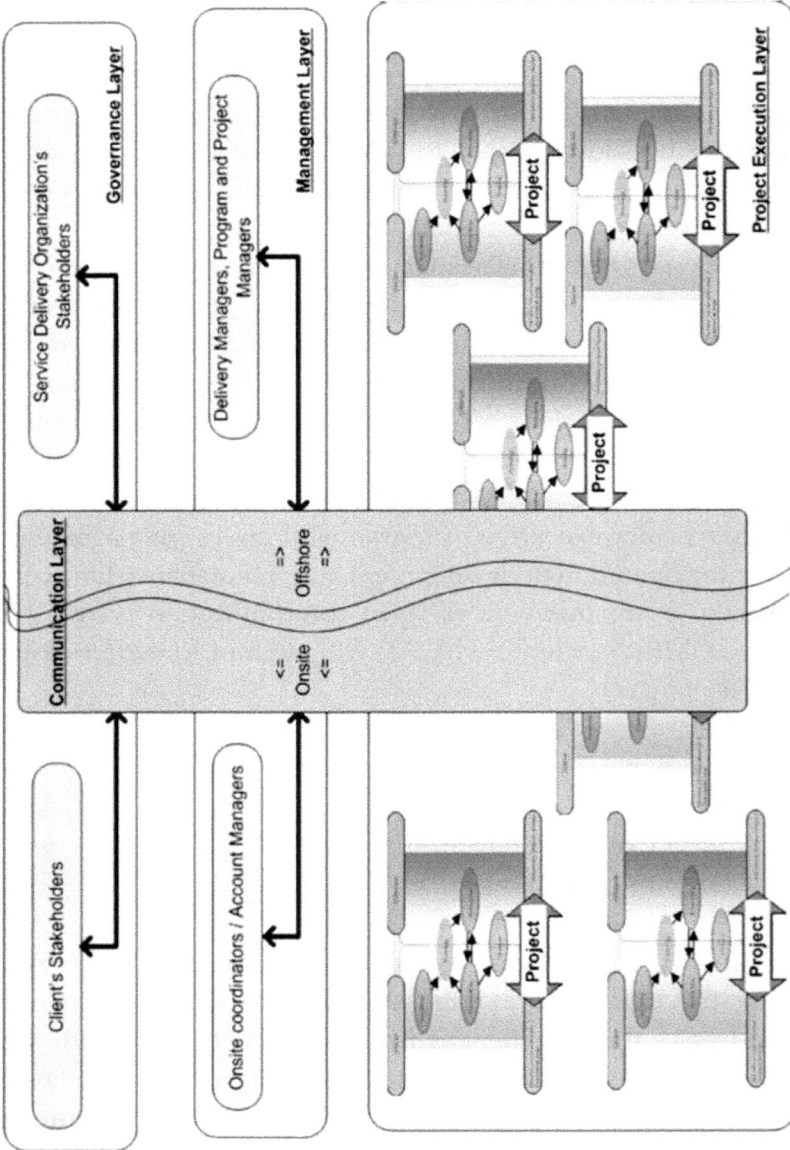

Fig. 3.1 Offshoring Management Framework

offshored. The Management layer extends the existing Body of knowledge on Project Management to address challenges of managing work efforts across geographic boundaries. Challenges of managing communication account for a large proportion of a manager's time and effort. It is then not surprising that tools and techniques of communication, including emerging technologies to facilitate remote collaboration and team development, are the highlight of any sourcing strategy. The Communication Layer of the Framework focuses on the challenges of communicating across geographic and cultural boundaries. Intricacies of managing the actual workflow of tasks, technology development and translating the business requirements will be highlighted in the Project Execution Layer of the Framework.

Offshore outsourcing is increasingly being viewed as a strategic practice of sourcing and managing business processes and software systems to low-cost offshore locations and includes Business Process Outsourcing (BPO) and Information Technology Outsourcing (ITO). However, the study of Offshoring Management Framework is primarily focused on offshore outsourcing of Information Technology and ITO. Managing IT projects and development is distinct from management of projects in other business verticals due to the strong *artistic engineering* focus that cannot be decoupled from the process orientation. Added to this is the challenge that stems from the debate over whether application software development is engineering or an art. Management of an engineering process is more process oriented and distinct from managing an artistic process. Interestingly, this debate is not really new; it was stirred over three-and-half-decades ago by Donald Knuth[4]. IT managers generally attempt to take a middle ground as they bridge the gap between the business and functional domains and technologists. The goal of managing IT Delivery projects is to ensure that the activities articulated in the Technical Domain segment (Fig. 3.2) conform to the business objectives on the left side and deliver applications and software solutions to address those objectives.

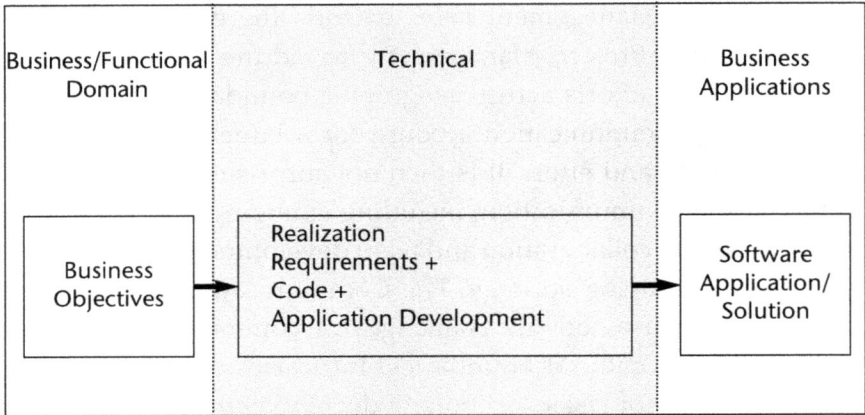

Fig. 3.2   Managing IT Delivery

The technical domain described in the figure above is perhaps the sweet spot—at the intersection of business needs and technical challenges—where Technology Managers add the maximum value. They attempt to marry their understanding of technologies and business drivers, while bringing the required management rigor and focus to the art and engineering of software application development. Getting technologists and business functional experts to communicate and orchestrate their diverging goals is a key challenge. Interestingly, the technical domain is also the key focus area in any outsourcing initiative. The realization of business objectives and requirements by ensuring that the application development efforts are synchronized towards building a software solution is perhaps the raison d'etre of technology offshoring.

GOVERNANCE LAYER

The governance of offshoring is increasingly being viewed as one of the most crucial aspects of a sourcing strategy. Executive sponsorship of offshoring initiatives is a key to successful development, delivery and deployment of projects and programs. '*Offshoring is strategically important and challenging—and requires strong governance. Most of the firms establish a steering committee from the very beginning to oversee initial deployment and on-going operations.*' says a recent Deloitte report[5]. Executive sponsorship is essential to facilitate global management and resolve issues and address challenges, and access to senior management will ensure that any escalations and issues are resolved before they get out of hand. The key dimensions of offshoring governance include:

1. **Definition of a Service Level Agreement:** This could be between an offshore and onsite team, or the client and the vendor.
2. **Program Management:** This could include establishing a formal offshoring Program Management Office.
3. **Transition Management:** Includes strategies and tactics for transitioning an organization towards becoming a strong offshoring player.

Offshoring governance teams will typically include executives from both the sourcing (onsite) organization and from the vendor (offshore) organization. Some of the common areas of focus for executives involved in offshore governance include:

- **Offshore Development Strategy:** Executives and managers of both the client and vendor organizations need to be clear about the strategic objectives of offshoring that could include reduced operating costs, shifting IT risk to a service provider, freeing IT management from operations management and staying ahead of the technology-adoption curve among others.
- **Steering Committee:** A steering committee with senior executives delegated from the client and vendor's organizations should be empowered to take strategic decisions and guide the operations of the offshoring initiative.

An example of an escalation to a steering committee could include contract rate changes. Though there may be a pre-existing Master Service Agreement (MSA) between the client and vendor regarding rates for certain types of services, the steering committee may be empowered to over-ride such rates if there is a sudden change in the market conditions. Such empowerment will help business continuity without impacting the overall performance.

- **Define the working model:** It is the responsibility of the governance council to ensure that the teams work towards building a flexible, collaborative work environment. Relationship and business goals change over time; therefore the executive focus should be on creating an environment that is adaptable to changes rather than on a view of the future based on the facts at a given point in time.
- **Articulating, measuring and monitoring the SLA:** A clear definition and understanding of the SLA by the governance council is essential to drive the expectations of the offshore and onsite teams.
- **Stakeholder Management:** Offshoring is a very sensitive issue, especially since it involves re-organization of the group/department during the transition phase and, in many cases impacts the lives of people in communities where an enterprise conducts business.
- **Dispute resolution:** Though all the stakeholders need to be optimistic about the success of sourcing, conflicts and dissent may appear from time to time. The governance council will also act as arbitrators of disputes arising out of individual contracts or performance of projects and pro-grams during a sourcing initiative.

Governance of an offshoring initiative includes aspects of managing, reviewing and monitoring the activities of teams and projects. Governance of offshoring includes transition management, and the management of steady state that may involve the typical challenges of executive management including benchmarking and ensuring that the projects and programs remain on track. It may also involve liaising with key stakeholders across the organization. Case in point is the offshoring strategy being adopted by General Electric[2] that is being questioned by a few shareholders. (Ref: Case in Point below) The bold and aggressive plan adopted by the company will constantly have to be justified to all stakeholders including the board and shareholders. This is one of the facets of stakeholder management that the offshoring governance council may be called in to address.

Box 3.2

CASE IN POINT: QUESTIONING GE's 70:70:70 RULE

The following is an extract from GE's Notice from the 2004 Annual Meeting:

Shareowner Proposal No. 5

The IUE-CWA Pension Fund, 1275 K Street, N.W., Suite 600, Washington, D.C. 20005–4064, has notified us that it intends to submit the following proposal at this year's meeting:

"Resolved: The Stockholders request that the Board of Directors establish an independent committee to: 1) prepare a report evaluating the risk of damage to GE's brand name and reputation in the United States as a result of the outsourcing and offshoring of both manufacturing and service work to other countries and 2) make copies available to shareholders upon request. "Statement of Support: In the 2002 Annual Report, GE announced targets of $5 billion in revenue from China and the outsourcing of $5 billion in contracts to Chinese vendors by 2005. China is a country where employees are persecuted for seeking to exercise internationally recognized human rights, such as freedom of association and the

Box 3.2

CASE IN POINT: CONTINUED...

right to collective bargaining. "GE is also attempting to outsource 70 percent of business processes (IT work, engineering, design, accountancy, legal services, call center work and bill paying), send 70 percent of outsourced processes offshore and give 70 percent of offshore outsourced processes to India. (Hindu Business Line, 6/18/03, reproduced on GE Capital India web site) India has been cited for non-enforcement of labor rights, including freedom of association and the right to collective bargaining. (ICFTU, 'Report for the WTO,' 2002)

"The outsourcing and offshoring of manufacturing and service work may be profitable in the short term, but could have significant long-term consequences. (Reuters, 10/31/03). The shift of production to low-wage countries in general and to China in particular has generated negative press stories in the U.S. (Knight Ridder, 11/10/03; Union Leader, 10/26/03) Two in three Americans think that job losses to China are a 'serious issue.' (Greenberg Quinlan Rosner Research, 2003)

"Americans are also sensitive to the exodus of jobs to India and other countries. (Time Magazine, 8/4/03) Observers predict a backlash against the outsourcing of white-collar jobs. (USA Today, 8/5/03; Business Week, 2/3/03) "GE is vulnerable to consumer disaffection in the U.S., which is the source of 60 percent of total company revenues. A backlash against outsourcing and offshoring could jeopardize political support for globalization, one of GE's five 'elements of growth.'

"Offshoring and outsourcing also affect the morale of employees who remain in U.S. operations. (CIO Magazine, 9/1/03) Morale problems extend to the countries where GE is sending work. A recent poll reported that GE Capital call center employees in India were dissatisfied with pressures to perform and insufficient time off. (India Business Insight, 9/30/03) "GE's brand name may be its most important asset. For Harris

Interactive, 'the value of a company's reputation may be as much as 40% of its total market value.' [http://www.harrisinteractive.com/pop_up/rq/benefits.asp] Company reputations affect consumer purchases. And 'reputation, once lost, is extremely difficult to reclaim.' [Wall Street Journal, 2/7/01]

"GE sends manufacturing and service work abroad. It uses foreign contractors. Its foreign operations are becoming vendors to other companies. We believe the Board should help shareholders evaluate the long-term risk and policy implications of the offshoring and outsourcing strategies."

SERVICE LEVEL AGREEMENT [SLA]

Offshore governance aims to establish the goals and expectations upfront, and one of the means of doing this is by defining and managing a robust SLA. SLA management essentially translates to the process of articulating metrics for tracking the progress of individual projects and programs and ensuring that they meet the goals of the overall systems. Effective governance of offshore IT management includes defining the business drivers and technical contract data. The business drivers could consist of the need to minimize financial losses and penalties due to service-level violations. It should also provide adequate incentives to perform above and beyond the metrics, which would translate to better ROI. The technical drivers include the need to improve productivity and lower the Total Cost of Operations (TCO) while enhancing best practices and overall efficiencies. Some of the aspects to consider while defining and managing SLAs of offshoring initiatives include:

- **Performance Measures:** Managers and project teams will need to benchmark and calibrate performance measures measurements to drive their tasks and to work towards achievable goals. The metrics to measure and monitor the business and technical drivers of projects and programs need to be clearly articulated in the SLA. The existing metrics, performance measures and processes will be considered while setting achievable targets and performance measures. For instance, if reduction in TCO is a measure defined in the SLA, the existing cost of operations and other financial measures will also have to be clearly articulated. Another example of a measure in a Batch Application system maintenance could be '% of times batch jobs were delayed.' The delay in percentage terms is measurable and easy to relate to. The SLA may have both positive and negative reinforcement; for instance, a positive reinforcement may be in the form of defining predefined targets and incentives for exceeding them. A negative reinforcement may be in the form of penalties for non-performance.

- **Scope of SLA:** In an offshoring context, the coverage of an SLA could range from that of individual projects to the scope of the entire engagement. The coverage and scope needs to be clearly defined and agreed upon by all parties. In large outsourcing engagements, the client and vendor may agree on different SLA measures for different work activities including Production Support and Maintenance, Application Development, Testing among others. Clearly defined scope will also determine the performance measures and help in monitoring the progress of tasks and activities. The SLA document may be an addendum to a formal contract agreement and may take on a legal slant as it attempts to set the agenda for sourcing and expectations along with the necessary due diligence.

- **Achievable Goals and Targets:** The goals and targets defined in the SLA should be achievable and realistic in order to provide for a 'win-win' agreement among all parties. The team members of service vendor and offshore delivery organization should be able to understand and work towards the goals. In certain instances, business goals may need to be translated into technical goals for the application systems since they may not be easily understood by techies. For instance, a business goal like 'achieving high customer satisfaction metrics' may need to be translated to measurable technical targets like 'Near Zero System Downtime,' '% of times response occurred within 15 min for critical items'. While setting goals and targets, typical SLA's will also call out causes beyond reasonable control, a.k.a. Force Majeure events. Typically, Force Majeure clauses cover natural disasters, "Acts of God" or war.

SLA definition and management is receiving increasing attention among business leaders and management experts who are working to understand and articulate the best practices of offshore governance. Thomas Lynch[6] articulates five key areas of focus to avoid pitfalls of offshoring, that include Basic contract terms and conditions, Agreement issues, International issues, Sarbanes-Oxley issues and Privacy issues. A well defined SLA with rewards for high performance and penalties for non-performance along with articulation of clearly measurable and achievable targets, limiting the responsibilities tied to an offshoring contract is an essential constituent of offshore outsourcing Governance.

TRANSITIONING OFFSHORING

Managing the transition from existing IT infrastructure to offshore teams is an emerging area of focus. Executives and managers at offshoring organizations need to be aware of the intricacies involved in moving from the initial, observer stage to a stage where they reap the benefits of offshoring. Organizations typically move through four stages of transition (Fig. 3.3) from observers to strong offshoring players. This also implies that translating an offshoring strategy to operations is not a 'project' that can be executed instantly. An offshoring strategy may simply be the desire of the senior management or a directive to move IT systems offshore. Translating that desire to fruition may involve intricate operational transformation that could take anywhere from 24 to 60 months. The real ROI may not be apparent till the *committing* stage or till the firm becomes a strong offshoring player.

The four stages of offshoring transformation are as follows:

1. **Observer Stage:** The first stage of sourcing maturity at an enterprise begins with the recognition of the need for offshoring; offshoring management needs to become an integral part of an IT strategy. During the initial *observer* stage, there may be very little traction in offshoring and the activities will involve gathering data and metrics and benchmarking the practices in the industry. At this point, there may be a need to define a business case for a sourcing strategy and observing trends in the marketplace.

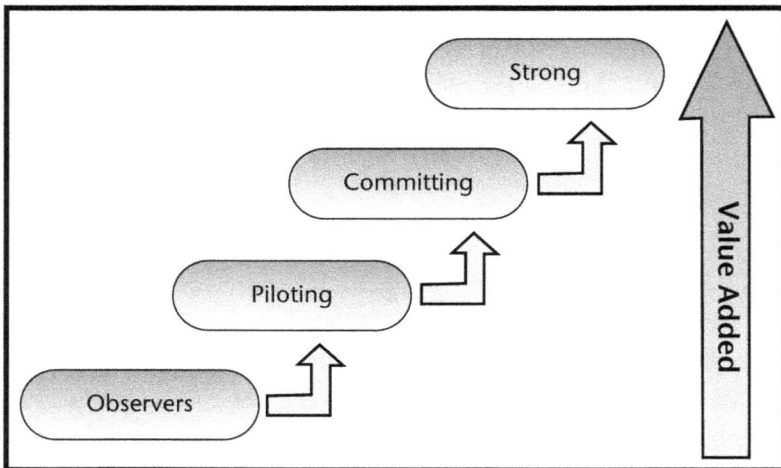

Fig. 3.3 Offshoring Transition

Organizations may solicit inputs on trends from software service organizations, consultants and analysts during the initial evaluation. Such evaluation will also set the ground for piloting applications for sourcing.

2. **Piloting Stage:** On recognizing the benefits of offshoring, organizations will graduate towards the *piloting* phase when they may identify one or more projects to be offshored. Alongside piloting, managers may also begin an application portfolio analysis, define the offshoring roadmap and select the appropriate sourcing models and vendors. The positive results from such piloting activities may convince the senior management to commit to an offshoring strategy.

   The offshoring roadmap may describe the portfolio of applications while giving weightage to individual applications that may derive maximum benefits in the least amount of time. For larger enterprises, the piloting stage may involve defining the overall sourcing roadmap including considerations of globalization, external landscape analysis and offshore vendor visits. Vendor visits are gaining significance as organizations contemplate larger offshoring engagements. Such visits may also include a tour of the country, detailed meetings and an evaluation based on a formal agenda.

3. **Committing Stage:** Organizations will be ready to commit themselves to offshoring after successfully piloting projects and defining an offshoring roadmap. This may include evolving governance techniques, offshore program management strategies and moving more complex applications offshore. This stage is also sometimes called the beginning of a *steady state* of offshoring, which translates to an opportunity for service providers and consultants.

4. **Strong Stage:** The end goal of offshoring strategies is to move to a stage of maturity where the organization optimizes on the benefits of sourcing. Working with vendors and executing projects with consistent predictability and

processes will help organizations derive the promised benefits of offshoring.

The level of maturity of an enterprise, along with the stage of sourcing, will influence the selection of the offshoring model and the transition into a mature offshoring stage. Management initiatives typically take off when business leaders and executives observe trends in published journals, articles or while they are involved in planning and benchmarking against competitors in the industry. Even though most large organizations are examining their offshoring strategies, few have transitioned to a steady state of offshoring.

The movement of an organization from the observer stage to becoming a strong offshoring player can be challenging and rewarding. Managing the steady state of offshoring may include aspects of general management, project management and also cross cultural management. During the committing phase, a firm may commit to a particular offshoring model and begin working with offshore teams, either its own or a vendor's. During this phase, the management will also begin documenting the benefits of sourcing and continue tracking the progress of sourcing according to the defined roadmap.

MANAGING OFFSHORING PROGRAMS

Projects, with fixed deliverables are the smallest units of managing discrete tasks and services that need to fit into an organization's technology roadmap. Technology initiatives are either undertaken as a fallout of strategic decisions to transform business units or to ensure that the business operates optimally in a steady state. They may be undertaken at strategic levels to ensure business transformation or to ensure maintenance and sustenance of applications where IT teams work towards ensuring that business systems continue to serve the changing business operations.

The technology initiatives often involve multiple projects aimed towards predefined targets that need to be managed in tandem with the overall goals and synchronized with the other projects being executed. This umbrella process of getting multiple projects to work towards strategic goals is also called program management. Program management aims to extend the traditional project management methodologies to synchronize the work of multiple projects, sometimes involving multiple vendors. In case of offshoring, the programs and global projects may span across geographic and cultural boundaries.

Multiple projects at large organizations are typically overseen by dedicated program managers who are a part of a formal group called a Program Management Office, sometimes called a Project Management Office (PMO). The groups or individuals are tasked with monitoring the activities of individual projects and bringing structure and formalism while buffering the technical teams from the issues pertaining to management and administration. PMOs typically work across both horizontal (business and IT functions) and vertical (top and middle level management and project teams) boundaries of an organization and in case of offshoring, even outside the organizational boundaries. The goal of a PMO is to provide tools, methodologies and resources to help managers streamline and automate many of the management tasks. The benefits include improving overall productivity and probability of success and more consistent and simplified management leading to higher customer and user satisfaction.

The office of Program Management acquires greater significance in an offshoring context. Organizations that typically source multiple projects to vendors need a single point of contact to manage and administer outsourcing programs. Such activities may be done by individual managers or a formal PMO. Vendors and service delivery organizations also dedicate Program Managers and Delivery Managers to act as a single point of contact overseeing multiple projects for clients. Some of the key areas of focus from a globalization context at PMOs (Ref: Program Management in Fig. 3.4) include:

Fig. 3.4 Program Management

- **Facilitating Communication:** Managing communication between teams and across projects, program and organizational boundaries is perhaps the biggest challenge managers face. Some of the communication challenges attributable to offshoring that PMOs focus on include:
  - ❑ Interfacing with the executive steering committee, especially on issues pertaining to contract administration, vendor negotiation and other third party management
  - ❑ Facilitating communication between teams, groups and organizations
  - ❑ Directing and synchronizing teams of people working across geographic and cultural boundaries

- **Managing Relationships:** The key to any successful off-shoring initiative is the management of relationships between the client and vendor teams, onsite and offshore. A Program Manager manages aspects pertaining to the relationship including negotiations, setting project expectations, analyzing the work output and other aspects pertaining to the smooth functioning of projects. Relationship management by an onsite PMO may include focus on liaising between individual business units, IT teams and offshore vendor teams. A PMO at a vendor organization will focus on co-ordinating with onsite teams and ensure that the different project teams work towards consistent, predictable delivery. This may include addressing project issues and any contingencies that may arise during execution.
- **Contract Administration:** Contract administration involves the process of ensuring that the vendor's performance meets pre-determined requirements. Contract administration, selection of approved vendors and relationship management, especially in large organizations is a specialized functional area. Contractor and vendor management can be a challenging activity even during the best of times and acquires a new dimension in case of offshoring projects. A sound contract is the key to a successful offshoring relationship and should encompass issues such as financial terms and payment, quality of service, escalation mechanisms, contract termination and closure, intellectual property etc. Contracts, by nature, are legally binding and enforceable. Managing contracts in an international setting can add new dimensions and complexities that Program Managers and administrators need to be aware of and include:
  - Vendor selection and management.
  - Monitoring, reviewing and approving billings, invoices and other financial aspects of contract administration.
  - Reporting, including verification of work tasks and activities, signing off milestones and other status reporting

    ❑ Analyzing Vendor performance and SLA. This may also include proactively suggesting corrective action where necessary.

- **Change Management:** Change management is an integral aspect of managing application development projects since changes in scope, requirement, technologies and other aspects could be in a state of constant flux. Global delivery of projects brings its own challenges and complexities including aspects of business changes, newer delivery models, emerging communication paradigms and best practices. It is the responsibility of the PMO and project managers to ensure that the changes are abstracted from the teams as much as possible. Also essential is the definition of guidelines for managing scope change and change requests. We will revisit aspects of change management in greater detail later in the book.

- **Infrastructure Management:** Offshored projects have their unique challenges and infrastructural needs. The challenges include the need for specialized software to manage communication and the hardware, network connectivity and environment set up and changes. Tools and technologies for remote communication and collaboration are essential for the smooth management of offshore development projects. Individual projects may have their unique needs that a Project Manager may need to address, but the common infrastructure like network, bandwidth, communication and other aspects may need to be overseen by a PMO.

- **Knowledge Management:** A PMO may also be expected to act as a repository of the project management best practices, points of reference for organizational data and documents, tools and templates. Large organizations typically have well defined tools and templates for most of the commonly encountered management processes and activities. Such automated tools encouraged unobtrusive information gathering and sharing. Program Managers should facilitate the use of such organizational tools and practices.

- **Travel, International Business Management:** In an off-shoring context, Program Management faces newer challenges pertaining to travel schedules and associated logistics including culture orientation of selected team members. The PMO also needs to be on the lookout for geopolitical changes that could impact projects and teams. This could include changing immigration and visa regulations in host and native countries, travel advisories and other aspects that could adversely impact trade and operations.

CONCLUSION

There could be many reasons why organizations consider offshoring. One reason could be due to benchmarking by executives and industry leaders who constantly look for emerging best practices in Technology Management. Offshoring being one of the emerging best practices, is being actively watched by managers across industry verticals and by academicians who are publishing an increasing number of articles and papers in journals. Offshoring initiatives, we observed, can be top-down or bottom-up. In the bottom up model, individual managers and groups in an organization begin to document success of small-scale offshoring organically before management decides to expand it across the organization.

There is a need for a vendor neutral offshoring model that organizations can use to develop their offshore outsourcing strategies independent of vendor's models. In this chapter, we introduced the Offshoring Management Framework, which is intended to provide a frame of reference for evaluating sourcing models. The OMF may also be adopted for use by vendors looking for industry best practices. The four major areas of focus in the OMF are Governance Layer, Management Layer, Project Execution Layer and Communication Layer.

NOTES

1. From the author's research notes of interviewing service providers.

2. General Electric [http://www.ge.com/en/]. GE's 70:70:70 rule, was originally articulated by its former Chairman, Mr Jack Welch. The vision goes on to say that 70 % of a business unit's processes should be outsourced. 70% of the outsourced processes should be offshore. 70% of the offshore outsourced work would be done in India. [Ref: http://www.ebstrategy.com/outsourcing/basics/definition.htm]

3. Cisco [http://www.Cisco.com] 'Cisco India rides high on outsourcing wave' [*The Hindu Businessline*, Jun 11, 2003], 'Cisco to outsource more work to India' *The Hindu Businessline*, Jan 13, 2003]

4. *The Art of Computer Programming* [Donald Knuth]

5. *The Titans Take Hold* [A Deloitte Research report]

6. Avoiding Pitfalls in International Outsourcing Agreements [Thomas J. Lynch, *Inside Supply Management*, September 2004]

CHAPTER 4

Offshoring: The IT Management Context

- The Management Layer
- Global Project Management
- General Body of Knowledge
- Organizational Practices and Tools
- Experience and Knowledge
- Globalization and Cultural Awareness
- The Global IT Manager
- Conclusion

The success of IT outsourcing lies in the consistent and predictable delivery of projects and products and includes reliable maintenance of applications. After the sourcing strategy has been defined and piloted, operational aspects including managing delivery and projects and co-ordination between offshore and onsite teams acquires greater significance. It is important to begin planning any outsourcing initiative by facilitating a strong project and program management process. Application of the available Body of Knowledge supplemented by an appreciation of globalization and management of disparate teams are perhaps the key success factors in any outsourcing initiative.

Management of Information Technology projects has received a lot of focus during the past decade as new development paradigms, software architectures and systems have emerged. The fast pace of change in technology along with speedier adoption by businesses has ensured the need for a faster turnover of technology application development. Individuals who are tasked with managing IT Projects, especially of Application Development and Maintenance (ADM) initiatives, have a discreet role to play in anchoring the effort to ensure delivery under pressure on time. They need to recognize the business drivers and ensure that the technology solutions help the business units and functional units perform optimally.

While focusing on the operational aspects, managers also need to keep an eye on technical innovations and breakthroughs that they can leverage. Project Managers take ownership of most aspects of software development, beginning with requirement gathering and culminating in a successful handover. The managers, however, do not work in isolation. They are supported by technical folks including architects, designers and developers who build systems based on several well established principles while adhering to a life-cycle model and adopting some of the techniques of software engineering.

THE MANAGEMENT LAYER

The Management Layer of the Offshoring Management Framework addresses the management imperatives of delivery and development of IT systems with teams onsite and offshore. Global managers draw inputs from various sources including general project management references, books and literature. There is also a wide array of tools and techniques available to them as they manage the development cycle. Figure 4.1 depicts the highlights of globalized projects, extending the basic management practices to address onsite and offshore management.

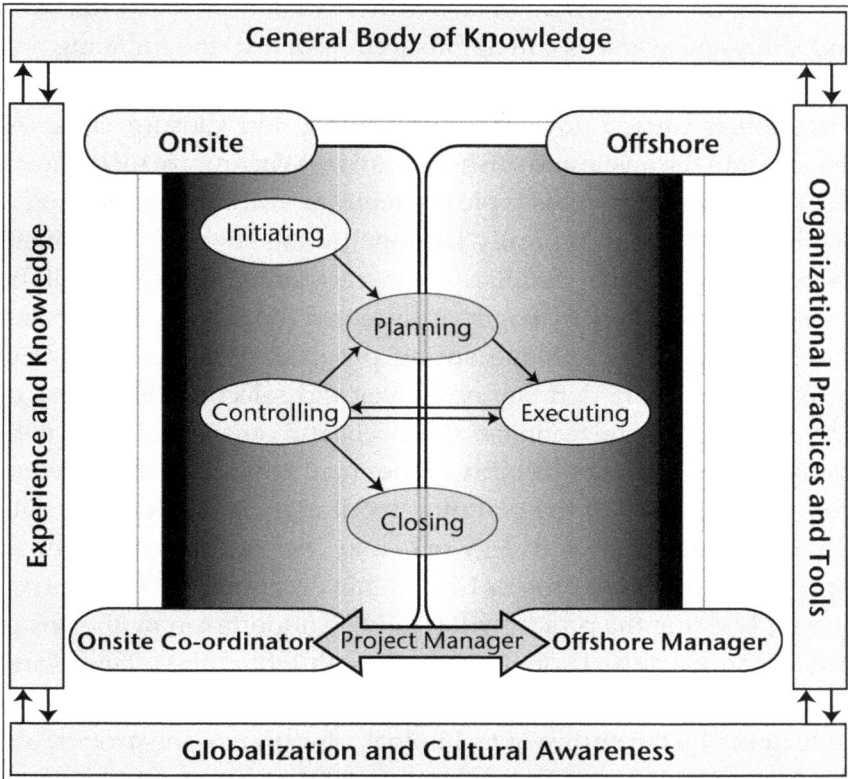

Fig. 4.1   Inputs for managing global projects

There are several popular project management processes and workflow mapping methodologies widely used in the industry. Among the more popular models is the description of the five key process groups in the Project Management Body of Knowledge[1] (PMBOK) by the Project Management Institute[1] (PMI). The guide highlights the dynamics of activities of the five process groups namely the initiating process, planning process, executing process, controlling process and the closing process. In our model we extend the processes into two main zones of operation, namely the onsite and the offshore zones. We will also examine the key inputs for Global Managers including aspects of the General Body of Knowledge, Organizational

Practices and Tools, Experience and Knowledge, and Globalization and Cultural Awareness. In an offshoring context, the Initiating and controlling processes are performed onsite, while the bulk of the execution is carried out offshore. Planning and Closure processes require both the onsite and offshore teams to synchronize their efforts.

The initiating process typically requires authorizing the project or phase which will typically be done by the sponsor or business owner based onsite. Similarly, the controlling process will be focused onsite where managers ensure that the objectives are being met by monitoring and measuring progress periodically. Project planning involving defining objectives and selecting the course of action, including the technical approach, and other details will generally be the joint responsibility of the client and the offshoring vendor's teams. Some of the planning work may be conducted onsite while the research and legwork may be conducted offshore. Similarly, the closure process of formally accepting the work products and closing the project will jointly be undertaken by the onsite and offshore teams. Though we have highlighted the offshore and onsite focus of activities, the actual separation of tasks may depend on factors like the nature of technology, duration of the project, offshoring maturity etc. The following are the key roles that may evolve in the management layer—both onsite and offshore—during project execution and delivery:

- **Onsite Activities**
  - ❑ *Onsite co-ordination:* Software service companies that execute offshore projects generally designate one or more onsite-co-ordinators to manage client interfacing activities. These individuals act as a liaison between the client's management and technologists and the offshore team. The client may also designate co-ordinators or supervisors from among its line-management to take ownership of specific projects or delivery initiatives.
  - ❑ *Onsite Account Management:* Onsite account managers take on the responsibility of managing multiple projects

for the client, acting as a single-point of contact for all escalations of project issues. They also manage the billing, financial and other contractual issues. At the client's end the account management may include billing and other administrative activities.

- **Offshore Activities**
  - ❏ *Offshore Delivery Management:* Delivery managers are responsible for co-ordinating with onsite account managers to ensure that project teams deliver the required outputs adhering to organizational standards and client requirements. Delivery management acquires greater significance in case of large accounts or at clients where the vendor is executing multiple projects; the offshore Project Managers may also take on delivery management roles in certain cases.
  - ❏ *Program and Project management:* Offshore Project Managers take the bottom-line responsibility of ensuring successful execution of projects. The term Program Management, though sometimes interchangeably used with Project Management, is generally used to denote the existence of larger projects managed by multiple managers.

Service delivery firms have designated roles of client facing teams including *Resident Project Managers, Onsite Co-ordinators* and *Engagement Managers* who are essentially responsible for the onsite-offshore co-ordination and delivery of work products and may also take on the responsibility of sales support, billing and other account management tasks. Clients and organizations sourcing IT work to offshore vendors may have well defined contractor management teams that are increasingly becoming offshore savvy. In many instances, IT managers at client organizations are also beginning to take on the responsibilities of managing global teams, especially as they begin to interact more with onsite co-ordinators and account managers.

Management of IT projects, even at the best of times, can be challenging. Statistics on project failure rates ranging from 50 to 70% abound in the industry folklore. Another interesting aspect of software development is the fact that the best (top 5% or so) programmers are about ten times better than average programmers. They tend to be logical and creative and can churn out solid code more efficiently than their peers and generally tend to bring out-of-the-box thinking into the programming process. Identifying and nurturing good programmers and teams and ensuring overall success of project delivery is a challenge. Figure 4.2 depicts a simplistic hierarchical structure at the lowest level: Project Managers work with teams of developers, technical specialists and architects. Most projects are initiated by articulating the different roles and responsibilities of individuals within projects. Project Managers manage the workflow and life cycle, Technical Architects and Module Leaders manage the technical aspects of development, senior developers may be assigned to design, manage unit testing, configuration and quality aspects, and

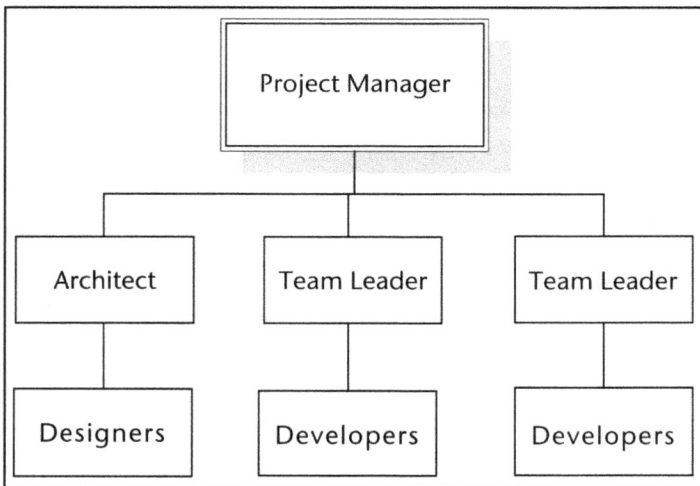

Fig. 4.2 Traditional Project Structure

developers and technical specialists may be assigned to design and develop individual modules.

Management of teams and the traditional reporting hierarchies are extended by offshoring. Adding offshoring to the technical and functional aspects of IT project management can exacerbate the challenges faced by managers. However, as the benefits of sourcing far outweigh the risks, managers need to learn to confront and address the challenges. The management of outsourced projects incorporates some of the best practices of general project management including planning, controlling, costing and scheduling, with a special emphasis on co-ordination between onsite and offshore teams. In addition, offshoring forces structure into projects and programs that may have been undertaken in an *ad hoc* manner, bringing a level of accountability and engineering focus to application development.

Figure 4.3, highlights a hierarchy similar to what we saw earlier with the additional focus on onsite/offshore development. The Project Manager continues to focus on the dynamics of managing technical teams, developers and architects where the bulk of development may be done offshore. In addition, a team onsite working with clients will have to be co-ordinated to synchronize the activities with the offshore team. The ratio of staffing onsite and offshore teams is a significant aspect of planning. Service delivery organizations attempt to staff teams in the ration of 70:30, skewed towards offshore development to leverage the low cost benefits of offshoring.

Project and program management in the context of an outsourcing initiative includes a group of people from both the client and vendor organizations tasked with ensuring that the planned effort stays on track. The actual composition of teams across geographic boundaries onsite and offshore may depend on a number of factors including the technologies, client-vendor relationship, risk tolerance etc. However, regardless of the compositions of the teams, the management of offshore and onsite teams encounter challenges that may follow a pattern.

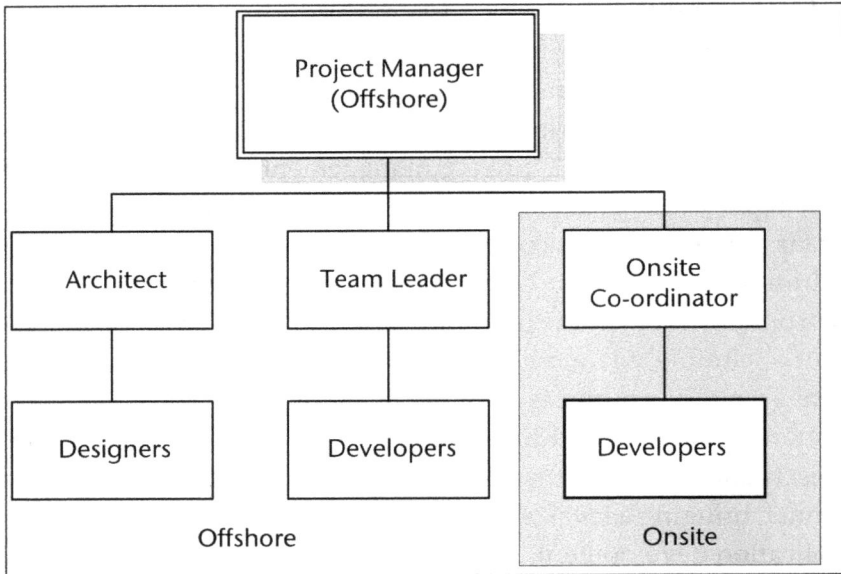

Fig. 4.3 Team Structure: Onsite/Offshore mix

GLOBAL PROJECT MANAGEMENT

We have looked at some aspects of managing global projects and the different roles onsite and offshore involved in synchronizing application development and delivery. This perhaps brings us to a basic question that most leaders planning and staffing for global teams grapple with: *who are global managers and what attributes should they possess?* Let us begin the discussion with a brief case in point, highlighting a recent advertisement.

Several key attributes stand out in the advertisement. Leadership of strategic sourcing along with a knowledge of IT Outsourcing models seems to be a key theme. Highlighted alongside are regular project management attributes including functional and business expertise and skills required to maintain client relationships and develop teams. Knowledge management and an

Box 4.1

CASE IN POINT: OUTSOURCING PROJECT MANAGERS WANTED

The following is an extract from a recent online advertisement from a leading organization calling for Offshore Project Managers.

Title: IT Outsourcing Project Manager/Analyst

Skills: IT, Project Manager, Manager, Analyst, consulting, project, program management, management, modeling, systems, analysis, lifecycle, development, security

Job Description: Work as a leader with the IT strategic sourcing team to provide clients with consulting services relating to organizational transformations. Manage functions, including providing strategies, solutions and support based on IT project management experience and knowledge of IT Outsource modeling. Oversee the proposal process, and provide assessments on acquisitions as it relates to company strategic plans. Maintain responsibility for providing functional, technical, and business expertise necessary to solve complex and strategic client problems. Develop intellectual capital and new service offerings, and support the growth and management of the business consistent with the IT strategic sourcing team. Serve as a recognized leader in the team, and act as a role model for the firm's core values. Develop staff and executive people programs across the team, and maintain client relationships. Manage business systems, including rate management, assignment, and contract management.

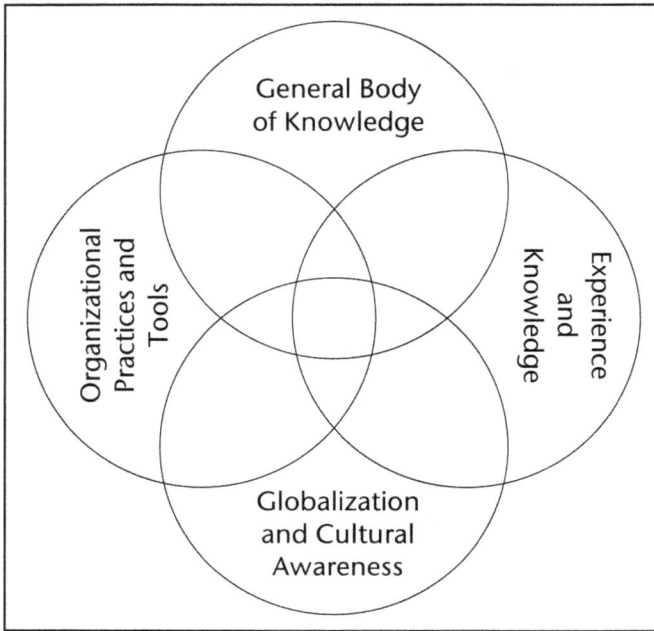

Fig. 4.4 Inputs for a Global Manager

understanding of organizational practices also seems to be a key attribute. In essence, a manager with the right grounding of technical and managerial skills and an understanding of the basics of communication and cross-cultural management is the key to the success of global projects.

Global managers draw inputs from various sources including the general project management reference, books and literature. Application development managers also have a wide array of tools and techniques available to them as they manage the development cycle. Figure 4.4 depicts some of the key sources of input. The factors described here are the same highlighted in Figure 4.1, albeit with a different view highlighting a tighter coupling. We will continue our discussion by examining the highlights of each broad area.

General Body of Knowledge

There exists a vast body of knowledge on general management that managers can tap in order to further their knowledge and reference while planning and executing projects. Universities and academic institutions regularly provide refresher courses on functional areas including finance, marketing, operations management, technology management, business strategy and, increasingly, on globalization. Books, magazines and journals dedicated to the field of study, along with inputs from professional bodies can also provide insight to managers. The Internet and web search engines are gaining popularity as tools of research and information gathering along with bulletin boards, discussion groups, newsgroups and other online forums that also act as bouncing boards for ideation.

Globalization, offshoring and outsourcing, along with emerging trends in internationalization, are entering the radar screens of management thinkers and academicians. There is a lot of literature currently emerging from both the business and academic publications on practices related to project management spanning global boundaries. Best practices are also emerging from the field and from service delivery organizations that are observing and developing practices that can be replicated elsewhere. Consulting organizations and analyst firms of various genera are beginning to provide offerings tailored towards offshoring including insightful whitepapers, templates and other reference collaterals.

Inputs from PMI's research is another source of reference for managers. The *PMBOK Guide* is a collection of knowledge areas pertaining to the different aspects of project management. (Ref: Box 4.2). The BOK is not specific to the best practices of Information Technology and is a reference for generic management practices that can span a wide array of domains and vertical areas where a structured management of project workflow is a key success factor. Trends and practices from offshoring management, co-ordination of onsite and offshore delivery and other intricacies are yet to extend into the formal BOK.

Box 4.2

PROJECT MANAGEMENT KNOWLEDGE AREAS FROM PMBOK

There are nine key areas covered in the *PMBOK Guide*, also known as "Project Management Knowledge Areas" that deal with the different aspects of managing projects and include:

- **Project Integration Management:** This is the stage where the project plan is described and signed off by the stakeholders. The other operational aspects of projects including planning of execution and change management and escalation procedures are also articulated. Among the key inputs that go into a successful project plan include planning and brainstorming, historical information on past projects, organizational policies and inputs from clients. Significant milestones, along with deliverables may also be included in a project plan. This is also the phase of the project where the key objectives of projects are articulated.

- **Managing Scope:** This activity is tightly coupled with the project planning described above and involves articulating all the work that will be required to be done to successfully complete a project. The Work Breakdown Structure (WBS) is one of the key activities performed while defining the scope. WBS is the breakdown of the project into tasks and activities to a sufficient level of granularity to enable a manager to assign individual tasks to team members. This breakdown is sometimes called decomposition.

- **Time Management:** It is essential that every manager focuses on timely completion of projects under a budget. There are several mathematical and project management software tools available to aid managers as they work towards timely completion. Techniques such as Critical Path Method (CPM), Program Evaluation and Review Technique (PERT) and other simulation techniques are extremely popular.

Box 4.2

PROJECT MANAGEMENT — CONTINUED...

- **Managing Cost:** This area of project management is closely linked to the Time Management described above, especially since the effort estimated generally drives the time and cost. An area of focus in cost management is budgeting and cost control that is significant for software services companies that not only have to keep an eye on the total cost of project but also ensure that they are deriving specified margins from the project initiatives.

- **Quality Management:** Quality assurance does not require extensive elaboration, but is an area in which many projects lack. As the projects progress and deadlines tighten, one of the easiest areas to cut corners is when it comes to quality, which is really sad since the entire effort of the project can be wiped away if the work is perceived to be of poor quality and does not meet the basic norms. Planning for quality assurance and control along with working with stakeholders on defining the 'cost of quality' is among the tasks of a manager.

- **Human Resource Management:** People are the key to successful execution. This is perhaps more true for the IT industry which hinges on the talents and skills of individuals. Ensuring that the project is staffed by individuals with optimal skills, motivating teams and creating and managing a cohesive team culture are among some of the key responsibilities of a manager.

- **Managing Communication:** Ensuring timely and appropriate generation, collection and dissemination of project information is a key focus area. Challenges of managing communication between individuals and teams may be magnified due to complexities of culture and geographies. Selection of appropriate tools and defining an acceptable mode of using them is one of the key success factors behind managing global communications.

PROJECT MANAGEMENT   CONTINUED...

- **Risk Management:** Projects will have to deal with an element of risk pertaining to different aspects. Managers need to identify, analyze and respond to real and perceived risks and plan for monitoring and controlling elements.

- **Procurement Management:** This area focuses on procurement planning, sourcing, contract administration etc. Although this is not a key focus area for software application development projects, managers may sometimes have to work on procuring software, hardware and other infrastructure. Managing logistics of international, local and regional travel is another aspect that may come to play in global projects.

Other specialized bodies of knowledge that managers regularly refer to include financial planning tools and techniques including cost/benefit analysis, Return on Investment (ROI) tools. They also get inputs from other functional areas and groups including Human Resources Management. Managers also rely on Estimation techniques and tools. In the software engineering world, there are also several formal estimation methodologies like Constructive Cost Model (COCOMO), Functional Point (FP) or Use Case Points (UCP) that can be used to get more accurate estimates. Estimations also play a significant role in negotiating the resources, timelines and budgets for projects.

ORGANIZATIONAL PRACTICES AND TOOLS

Most large software and service organizations have strongly defined proprietary processes and tools that they expect their managers to be conversant with. Some companies even use such proprietary processes as a differentiator in the marketplace, the processes also play a key role in enabling managers. Use of the right tools and methodologies ensures consistent delivery and helps faster and better integration with the systems being developed. Identifying the right tool or mix of toolkits along with adequate training and access to best practices is also essential to consistently deliver quality output. Organizations use tools and templates for different aspects of project management including:

- **Project Planning and Tracking:** Ensuring that the different milestones of projects are successfully completed and delivered to the customer is a manager's responsibility; along with that they also need to ensure that the client is billed for the work delivered and that the (net profit) margins accruing to the organization are on target. Tools for time and effort tracking including time entry, tracking schedules and milestone can either be custom built or acquired commercially off the shelf. There are several project tracking tools in the market including products like Microsoft Project, etc. Many organizations also develop custom workflows over such commercial products to address specific planning and tracking needs.
- **Contract Management:** Service companies depend on market intelligence and experiences to help win contracts and bids. In order to manage large portfolios of projects they need sophisticated tools to manage project proposals, opportunities and the workflow involved in such activities. Client organizations also integrate sophisticated contract management software into their financial reporting and management systems.
- **Tools for Managing Communication, Workflow etc.:** The key challenge in managing projects is in ensuring consistent, clear communication across the organization and teams onsite and offshore. Emails, instant messengers, blogs and other

internet based tools have gained popularity during the past few years. Responsible use of such tools is a key management imperative.

- **Templates and Historical Data:**   Large organizations have an extensive collection of templates for requirement gathering and for the other aspects of the software Life Cycle. Mature organizations also build an extensive repository of historic data based on the experiences of their consultants. Such historic data helps in planning, forecasting and in developing solutions faster and better.

- **Quality Assurance:**   Automated testing, validation and verification tools can aid productivity and help deliver quality code. Tools can also aid in bug reporting, defect tracking, issue tracking, scripting and test automation. Configuration and code management, version and source control is another area of opportunity for the adoption of right tools.

- **Globalization and Cross Cultural Management:**   Managing globalized IT projects needs an awareness of trends in the global marketplace with a grounding in managing cross cultural aspects of geographically dispersed teams. We will examine aspects of team management in global settings in further detail later in the book.

Tools and templates help organizations capture the best practices of past projects which in turn act as a point of reference for future projects and facilitate the speedy, consistent gathering of requirements. Templates are especially useful for service organizations that regularly capture requirements from clients across domains and verticals. For instance, the requirements of a healthcare customer may be totally different from that of a financial organization. The use of templates will help analysts and managers address the key issues that most projects may face. Along with the tools, best practices in the form of design patterns, architectural styles, templates and BOK are also widely used and advocated.

Box 4.3

CASE IN POINT: INFOSYS[2] TOOLS AND PROCESSES

Infosys' processes and in-house tools help managers and project teams work towards consistent end-to-end delivery. Though I have a good understanding of the processes and practices at Infosys, it will not be fair for me to describe it since Pankaj Jalote[3] has already articulated some of the key processes and methodologies at the company including the Process Database (PDB), the contents of the Process Capability Baseline (PCB), Process Assets and BOK system. The book explains the project process along with a highlight of the organization-wide process and includes sections on Requirement Change Management and Effort Estimation and Scheduling and also talks about Infosys' Quality Planning, Metrics and Statistical Process Control (SPC). Some of the key highlights of Infosys' tools and processes to facilitate its proprietary Global Delivery Model (GDM) include:

- Strong quality and project management processes that ensure consistent delivery
- World class knowledge of management practices and systems that facilitate knowledge sharing and cross-pollination of ideas among teams.
- Processes for interaction and communication within teams make it possible for globally distributed groups to interface and collaborate seamlessly
- Tools that monitor projects to track defects and benchmark them against estimates
- Tools, such as Influx, for scoping, requirements gathering and impact analysis
- Tools to monitor efforts, schedule adherence and slippages
- Process assets systems and tools to efficiently store and manage project documents and data
- Specialized tools to track individual service projects like application maintenance

Box 4.3

CASE IN POINT: CONTINUED...

- Tools to monitor automatic scheduling of audits based on detailed guidelines, followed by tracking of audit results, non-conformance reports and corrective actions
- Tools, such as PRISM, to automate the workflow for senior management reviews, in line with engagement schedules and plans

The above discussion was intended to highlight some of the major areas where tools and templates are essential for the success of execution. This is by no means an exhaustive list and is merely indicative of the wide array of tools and best practices available. We will continue to examine the use of tools and technologies and other aspects of managing global projects in further detail in forthcoming sections.

Experience and Knowledge

Mentoring and learning from peers is another valuable technique to build strong management skills. Mentoring helps foster the organizational culture and builds an environment of knowledge sharing since inexperienced employees are not afraid to make mistakes when being watched by their experienced colleagues. Executives also realize that formal and informal mentoring in organizations is one of the most efficient and cost-effective ways of achieving corporate growth.

Large software service organizations employ dedicated teams of individuals who benchmark the best practices in project management, review tools and recommend them for internal use. They also pilot such tools and techniques and assist in training and enablement of project teams. Project Managers at such service

Box 4.4

Case in POint: Knowledge Sharing at Infosys

Organizations in highly dynamic and innovation-focused industries such as telecommunications and networking have long used
Knowledge systems and repositories. KShop (Knowledge Shop
portal) at Infosys is one of the many innovative tools the organization uses to facilitate greater reuse of the best practices existing
in pockets. Based on principles of Knowledge Management, the
system reduces risk and helps build the robustness necessary to
thrive in a changing environment. KShop serves about 20,000
requests a day, translating to an average of about one document
from the portal reused every two work minutes. Content of various formats from sources across the organization are vetted, peer
reviewed and added to the database. Another focus of KShop has
been to minimize the overhead associated with creating content.
For example, KShop generates project snapshots on the fly from
existing Infosys project databases, thus minimizing the need for
manual compilation of these snapshots. By institutionalizing best
practices existing in pockets, facilitating greater reuse and
helping better virtual teamwork, Knowledge Management at
Infosys raises the ability to deliver greater quality and achieve
faster time-to-market. Within a relationship, knowledge management processes operate at three levels:

1. **Project Level:**   Teams have a project management co-
   ordinator for each project and specific knowledge management related goals within projects. Periodic project reviews
   cover project management as well.
2. **Account Level:**   Customer Accounts at the company have
   a knowledge management roadmap drawn out for them.
   Teams within accounts draw heavily on the Infosys
   Knowledge management systems. Knowledge sharing is
   strengthened by a range of methods such as orientation
   training programs, online discussion boards and collaborative environments within projects.

Box 4.4

CASE IN POINT: CONTINUED...

---

3. **Organization Level:** The KShop portal hosted on the intranet encourages organization-wide knowledge sharing and management ethos. This also ensures that the teams have access to the best practices and the collected learning from client organizations.

    The company recently joined the elite group of Knowledge focused organizations by appearing in the Most Admired Knowledge Enterprise (MAKE) listings. Others who have made the list include technology focused organizations such as Lucent, Nokia, Cisco and British Telecom.

---

organizations have the benefit of gaining from experiential learning of their peers and colleagues. While managers at smaller service providing organizations may not have the same edge in terms of formal training and access to tools and frameworks, there exists a sufficient body of knowledge, published case studies and reference material capturing the best practices across the industry.

Globalization and Cultural Awareness

IT Project Managers around the world are coming to realize the impact of globalization in offshoring development and maintenance operations. It is increasingly becoming significant for managers to be aware of trends in the global marketplace, along with a good grounding on aspects of managing cross-cultural and geographically dispersed teams. Managing global teams also involve emphasis on cultural awareness, nuances of language, communication mode and use of technologies to facilitate remote communication. The offshore application development model hinges on reducing costs by managing people in different corners of the globe. This also means that project plans will attempt to minimize cross-country travel during the course of the development life cycle. In order to effectively manage geographically distributed teams, Project Managers need to be extremely empathic towards members of team who are remotely located.

Box 4.5

CASE IN POINT

During my stint at a telecommunications company in Colorado, I had an opportunity to work in a multicultural and multinational team, though we didn't really formally call out the multinational aspect. This was the height of the dot.com era and our company had sponsored H1 visas for nationals from several countries. There were a few Americans including our business unit manager, a couple of Indian and Korean consultants, an Australian and a Canadian. Although everyone on the team was fluent in English, the accent and colloquiums used by individuals took a while to get used to. Our team meetings would invariably lead to a few interesting side-discussions and friendly jibing on news and happenings 'back home,' be it India loosing a Test Cricket match to Pakistan or Australian tourists stranded in England and the like. The manager very effectively used such jibing sessions to create a sense of team camaraderie and ensured that those of us from 'collectivist cultures' (Indians and Koreans) did not feel left behind. As the months progressed, the team began to work as a cohesive unit and produced stellar output without the use of sophisticated 'tools and techniques' of international business management and culture.

THE GLOBAL IT MANAGER

Organizations continue to build IT project management competencies and have also begun to realize the need for a cadre of *global* managers who can manage offshored and outsourced projects. Service Delivery organizations are also scaling up their global project management competencies and are grooming managers to take on such responsibilities. There are several checklists for selecting effective global managers published by international business gurus. A case in point is Matsushita's SMILE selection technique that stands for Specialty, Management ability, International, Language, Endeavor resonates well with international business planners. Similarly, Erran Carmel[4] articulates the unique qualities with the acronym MERIT, unique qualities that allow the global software manager to handle multicultural and dispersed components. The five unique qualities are Multiculturalist, E-Facilitator, Recognition promoter, Internationalist and Traveler. Successful global organizations take a base framework from the published literature and build aspects suitable to their specific needs on top of it based on their organizational culture and the global context in which they operate. It also follows that managers of global IT projects need to wear multiple hats depending on the circumstance and exact nature of the projects being managed. Apart from project management and technical knowledge, a global manager should possess key traits that include:

- **Project Management Skills:** Global IT managers are first Project Managers since they deal with projects and programs that in turn form a part of the overall offshoring and globalization strategy. In the previous section, we examined some of the key inputs for global project management including the general body of knowledge, organizational practices and tools, experience and knowledge and globalization and cultural awareness.
- **Strong Communication Skills:** Managers spend a lot of time communicating with their team members, across

teams and vertically with senior management reporting status and ensuring sponsorship. The need for a global manager to possess strong written and oral communication cannot be understated.

- **Technical and Domain Knowledge:** A good part of a manager's job involves managing aspects pertaining to communication across and outside teams. Knowledge of the technologies and business problems acquires a greater significance in a globalization context where impediments like time, culture, language and accents can creep into communication; if the manager can at least assert a common ground in terms of the technologies and business domains, the process of communication will be more streamlined.

- **Open to Travel:** Travel across geographies is an essential aspect of managing global teams. A manager may be expected to spend considerable time on the road to ensure 'face time' with teams. The manager should also facilitate travel of team members and be on top of other logistics including being updated on visa regulations, travel advisories, guidelines and governmental regulations etc. Interestingly, motivating techies to travel is a topic that I covered in a column of mine (Ref: Box 4.6)

- **Cultural Sensitivity:** Global managers need to have a good understanding of the aspects of cultural differences and sub-tleties of communicating across cultures. This is a significant area of focus since the success of communication in a global context will hinge on the rapport between parties communicating; and such a rapport will flow from mutual understanding of the individuals who may be from different ethnic, cultural and regional backgrounds. Managers can draw on many sources of information including international business management practices while preparing themselves to manage globalized teams.

- **Outsourcing Experience:** Technologists who have experienced working with onsite and offshore teams during the

course of their work may already have an understandin;
some of the fundamentals of teamwork across cultι
boundaries. Many of them may take on client-interfac
roles and travel to client locations during the course
project execution. Such experiences equip them with
skills required to interface with peers from other cultι
and ethnic backgrounds. By coupling the experier
learning's reinforced by formal training such individι
may be good candidates to take on management of glc
IT delivery.

- **Updated on Geopolitical Trends:**   This is a trait that sc
international management experts also refer to as 'inteι
tionalist.' Environmental, governmental, political and oι
external events in the business landscape can directlɣ
indirectly impact projects. By being aware of change:
the marketplace and the global geopolitical environmι
managers will be in a better position to anticipate and ρ
for risks, empathize with team members from diffeι
regions and help facilitate teamwork.

In this section we examined some of the key attribute:
offshoring managers. Earlier in our discussion on offshoring,
examined the dynamics of managing offshored projects,
yin-yang between offshore and onsite teams and between client ;
the sourcing organization. Project Managers may be appointec
either *side* of the offshoring spectrum to manage aspects of a prc
or delivery of a vendor's projects. While focusing on the project ι
on hand, managers may have to step up and work towards the
picture. This may mean that a manager at a vendor organizaι
working for a client's project needs to take on the responsibiliι
being an ambassador of the client; likewise, a client's manε
should empathize with the vendor's delivery team and facili
them as much as possible.

Box 4.6

ARTICLE: OF TECHIES AND TRAVEL

Face-to-face interactions and travel for work still remains the prime focus in the field of technology consulting, even with the advent of tools and technologies of modern communication like cheap VoIP phones, videoconferencing, etc. During the mid-nineties, before the advent of widespread outsourcing and 'global delivery,' staff-supplementation, that is body-shopping was the most popular model. Companies that wished to augment their IT workforce would contract a body-shop that would, literally, go halfway across the world to look for suitable candidates, process their visa and paperwork and make sure they landed at the client's place.

Towards the end of the nineties, management and technology gurus began to predict a total shift in paradigm with the advent of newer technologies to enable remote meetings and communication. This theory got a boost with tightening immigration and security laws in the West after 9/11. The current trend of global outsourcing and geo-political changes notwithstanding, travel continues to be an integral part of a typical IT worker's landscape. Though travel is not as widespread as in the mid-nineties, Indian professionals continue to crisscross the globe.

The glamour of travel to exotic lands aside, there is a human angle to all the travel. Let us take two extreme cases. The first involves a mid-level guy, let's call him Raj. He has been angling for a foreign trip for a while and was elated when his manager asked him to get ready to travel to Canada for a fortnight's technical requirement analysis for a client's project. He throws a party for his friends, packs his suitcase and heads for the airport with his family there to see him off on his maiden foreign trip. After Raj bids adieu to his folks and is ready to check-in, there is an announcement on the public address

Box 4.6

ARTICLE: CONTINUED...

system, asking him to report to the customer relations officer who says there is a message from his company asking to call their hotline. On calling the number, he is told that the client has shelved the project and he has to scuttle his travel plans. Raj is mortified by this turn of events and drives back home with his family.

Another case is that of a manager, let's call him Kumar, who manages three projects out of a multinational company's offshore centre in Bangalore. Having been in the industry for nearly a decade, Kumar has literally been-there, seen-it, and is not exactly keen on getting yet another immigration check stamped on his passport. Kumar also has a few personal issues, including the expected addition to his family, because of which he is not excited by the prospect of traveling. However, the 'problem' is that he is one of the few people in his division to hold a 'coveted' US H1-B visa. Every time an onsite requirement comes up, his bosses look to him. After dodging the bullet a couple of times, Kumar feels that he is really under the gun and is torn between his personal obligations and the expectations of his employer.

There is one common thread running through both the stories that travel remains a highly contentious issue, beyond the control of most individuals. Though individual developers, architects and managers remain in control over most aspects of their work lives and careers, aspects related to travel remain out of their control. There are probably several reasons for this. Most travel requirements are driven by a client's needs. Travel is generally billed to individual projects, in turn paid for by the client. To further complicate the matter, are issues related to international travel, including visas and immigration control. It is not surprising to find individual techies bemoaning their total lack of control over travel.

BOX 4.6

ARTICLE: CONTINUED...

I posed a question on issues related to travel to several managers at Indian companies known to me and most of them just shrugged their shoulders and accepted these anecdotes as a way of life for techies. Many also indicated that larger software houses are beginning to make serious effort to mitigate the need for constant travel by prodding wider adoption of technologies including video, voice and teleconferencing.

They, however, also conceded that adoption of such technologies is still at a nascent stage. Individuals and clients still seem to prefer the comfort of an eye-to-eye meeting and the 'touchy huggy feeling' of shaking a hand and explaining a system problem to an architect and to see him/her design the system.

Till more of us technocrats and managers begin to push for adoption of remote meeting technologies, the Kumars and Rajs will continue to be on tenterhooks.

(Originally published as a column in IT People section of Express Computers)

CONCLUSION

In this chapter, we examined some of the intricacies of the Management Layer of OMF and the three key areas of input for Global Project Managers including the PMBOK, organizational tools and techniques and the significance of continuous learning and knowledge building. Managers draw from myriad sources as they manage cycles of projects with increasing complexities. Managers of offshored IT projects should be comfortable in managing technology and possess strong technical and project management skills along with an understanding of general management practices and techniques. In addition, they should have strong oral and written communication skills, contract negotiation and management background, organizational skills and an awareness of the nuances of offshoring. In the rest of this book, we will continue to examine key tools and techniques with an emphasis on managing IT service delivery projects in a global environment.

NOTES

1. The Project Management Institute (PMI®), a premier professional body dedicated to the study of emerging practices, has developed an extensive collection of best practices, also called *A Guide to the Project Management Body of Knowledge (PMBOK® Guide)*.

2. Infosys [http://www.Infosys.com]

3. *Software Project Management in Practice*, Pankaj Jalote

4. *Global Software Teams, Collaborating across borders and time zones* [Author: Erran Carmel, PHI]

   General: In this discussion, and in the rest of the book, the term "client" or "user" is used interchangeably to imply the end user or the client of the service delivery organization. Vendor is a term we will use interchangeably with the Service Delivery organization. Depending on the offshoring model, especially in case of subsidiary and Joint Venture models, the 'client' and 'vendor' may be different divisions in the same organization.

CHAPTER 5

Project Execution Layer

- 🖥 Planning
- 🖥 Controlling and Monitoring
- 🖥 Closing
- 🖥 Change Management
- 🖥 Quality
- 🖥 Customer-Vendor Relationship Focus
- 🖥 The Offshoring Sweet Spot: Project Execution
- 🖥 Conclusion

The Project Execution Layer of the OMF addresses facets of an off-shoring operation. Offshoring decisions, planning and management are generally done by senior managers and business executives. After the offshoring strategy is defined, and the offshoring management practices and infrastructure are integrated, the focus shifts to individual projects. In a sense, the Project Execution Layer is the most significant layer of the framework since it addresses the tactical aspects pertaining to individual projects and work executed at the lowest level of granularity of offshoring involving collaboration between onsite and offshore teams. Depending on the client and context, other formal names for project execution include *Work Order or Statement of Work*.

The process group interactions highlight the dynamics of project life-cycles during execution. (Figure 5.1) The five key process groups highlighted in the figure extend from the Project Management Body of Knowledge[1] (PMBOK) that managers and teams may already be comfortable practicing. In our model, we extend the processes into two main zones of operation to address the onsite-offshore model of offshoring. As depicted, initiation and controlling activities will generally be the responsibility of onsite teams and the execution (actual application development and coding) will primarily be concentrated offshore. The activities pertaining to planning and closure of projects will be conducted jointly by the onsite and offshore teams to synchronize their efforts. The initiating process typically requires authorizing of the project or phase which will typically be done by the sponsor or business owner based onsite. The actual separation of tasks onsite and offshore may depend on factors like the nature of technology, duration of the project and offshoring maturity.

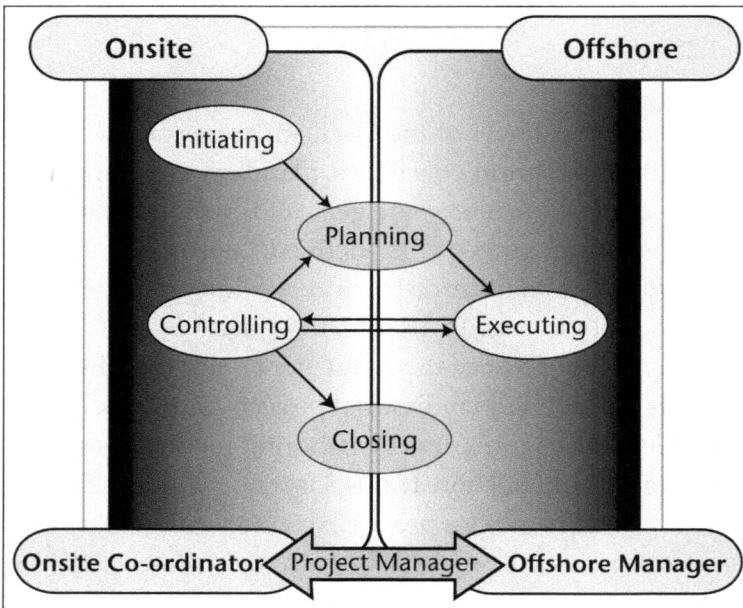

Fig. 5.1 Execution context

The management of service delivery projects draws extensively from the generally accepted global best practices and the Body of Knowledge. Project teams also take inputs from their organizational history and knowledge base while planning and executing projects. The life-cycle articulated in Project Management literature is generic enough to span projects across business domains. Planning and execution of offshored service delivery project takes off after a proposal has been accepted by the client, an agreement signed or an approval obtained from senior management. Some of the key areas of focus from an IT Service Delivery context include the aspects of Planning, Controlling and Monitoring, Requirement Gathering, Knowledge Acquisition, Change Management, Project Learning's, Quality Management and Customer-Vendor dynamics.

PLANNING

Planning involves thinking through and strategizing on all the aspects of a proposed project or initiative. The stakeholders and project sponsors generally specify the key objectives of the project that form part of the planning process along with formal requirement documentation and agreements or work-orders. Organizational policies and infrastructure including communication, staffing, quality assurance, budgeting, risk-response etc are also inputs for managers during planning. Other inputs to the planning activity include historical data and statistics on past projects, client preferences and requirements. Contractual agreements between the vendor/service provider organization and the client are also a major input used in planning. Planning also involves extensive negotiations and discussions with stakeholders, sponsors and customers, brainstorming meetings and forecasting. During planning, a manager needs to document key assumptions that can affect the course of the project along with the constraints. Typical constraints include a limited budget, timelines, availability of resources, access to subject matter experts etc. Inputs of the planning exercise feed into a Project Plan, sometimes known as the 'Master Project Plan'.

A Project Plan is a formal document that lists dates and timelines for performing activities that are scheduled to be delivered at different milestones. A plan is a living document that changes during the course of the project as more details about the tasks and activities emerge. As can be observed in Fig. 5.2, a project plan takes in inputs from a wide array of sources. For large projects, a Master Project Plan may involve planning and documenting subsidiary management plans.

The project execution planning is generally a joint exercise undertaken by onsite and offshore teams. Managers at large service delivery organizations may have an edge over their other peers since they will have access to extensive project metrics and repositories that document the actual project data from past engagements. Such historical data will help estimate the effort, plan staffing and assist in prescribing an offshore onsite mix that the team is comfortable with.

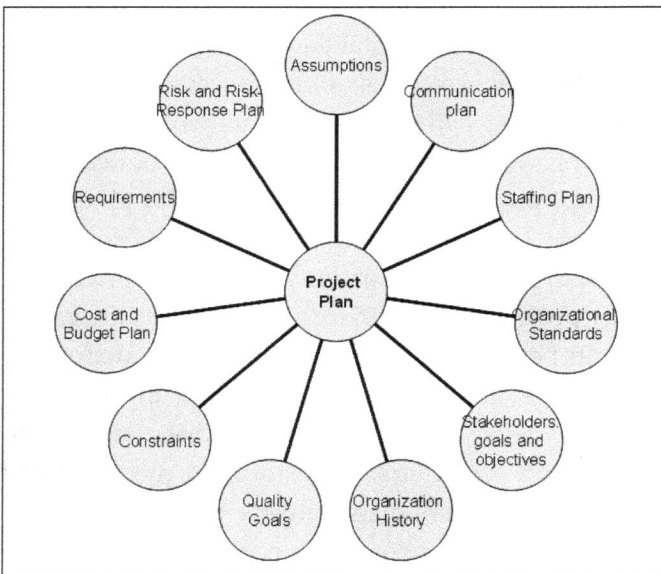

Fig. 5.2 Inputs to a Project Plan

Structured project plans can help reduce uncertainty and mitigate risks. Project plans are also effective communication aids that can facilitate in articulating the goals, objectives, tasks and activities to all the stakeholders, including clients, project sponsors and the project team. A properly documented plan will help managers think through the various outcomes and plan for necessary corrective actions.

CONTROLLING AND MONITORING

Monitoring of project activities commences right after the first project task begins and continues till the project deliverables are successfully handed over. Any deviation from the plan has to be monitored and corrective action suggested. By efficiently monitoring the landscape and work products of the team, a manager can initiate preventative action on tasks and activities that may slip in schedules, risking timelines and deliverables. This will also help managers set corrective actions on tasks that have over-run cost and budget. Special attention needs to be given to the risks documented in the project plan.

Monitoring the activities of project teams is an art more than a precise science. Software professionals are generally well educated and pride themselves in their state-of-the-art skills. They abhor being 'managed' and feel strongly about anyone peering over their shoulders. When managers begin to obtrusively gather inputs, they may be perceived to be controlling, an act that may lead techies to rebel. Another cultural trait common among technical folks is the common disdain for processes and controls that they associate with bureaucracy. Programmers and Software Engineers prefer to "just do it" rather than describe how they are going to solve problems, document and record them. Some of the practices in the field go against the grain of structured Software Engineering practices that dictate that all activities need to be thought through, designed, documented and then implemented.

Controlling and monitoring progress of offshored projects may generally be done onsite, though offshore teams may be involved in gathering metrics and reviewing the monitoring activities. Managing and monitoring the activities and outputs of technical professionals requires a lot of perseverance and tact. Educating team members about the significance of monitoring and control, and getting their active participation is perhaps the key to success. Systematic control and monitoring of activities in the project life cycle ensures:

- **Under-Budget, On-Time delivery:** One of the key goals of any project is to ensure delivery under budget and on time. Most clients have fixed budgets allocated for projects. The timelines for many initiatives also have strong business drivers.
- **Customer Satisfaction:** By ensuring delivery under budget and on time, onsite–offshore teams can win the confidence of customers. By doing so, the manager also ensures that the client builds a favorable opinion of the service delivery organization.

Every project grapples with issues pertaining to budgets, resources, people and slipping schedules and deadlines. Therefore, managers need to be extra vigilant in ensuring that they have a system in place to constantly monitor progress.

CLOSING

Project closure involves acceptance of deliverables by the client or project sponsors, and could include a formal sign-off or informal acceptance and use of deliverables. Most service delivery organizations encourage a formal closure and documentation of project metrics during the winding-down phase. There are several advantages to documenting methodical project closures which include:

- **Adding to the organization's project history and knowledge repositories:** Lessons learnt in the project including variances from project plans, corrective actions, risks faced etc., when documented become a part of an organization's project repository. Project history of past projects is also a key input used during project planning. Keeping a historical perspective on projects help organizations and managers gain insights akin to drivers using rearview mirrors to safely navigate through traffic.

- **Estimation:** Documenting and storing a project learnings in a well designed database helps organizations glean a lot of business intelligence that can be a source of competitive advantage. Knowing the actual timelines and effort involved in projects can help managers estimate future efforts with greater accuracy.

- **Individual learnings:** Individuals learn and hone their management skills by managing cycles of projects. During every execution, they face new challenges that help sharpen their skills. Project closure can be a time of introspection when they can reflect on 'what worked and what didn't work'.

- **Case studies:** *'Nothing sells like success'* is an old marketing adage. Successful project implementation, client references and kudos can be captured and showcased to prospective clients. If the project involved any new, innovative, path-breaking techniques, the workings of the project should be studied in greater detail to reverse-engineer the best practices.

- **Failure analysis:** Analyzing and sharing details of failures can be a good learning tool. Analysis of failure can help in identifying gaps and opportunities in processes and help in planning for staff training.

Although there are several benefits that accrue on methodical documenting of a project's learnings and best practices, this is harder to enforce since projects regularly schedule release of developers and other resources once the user acceptance test commences. Closure of projects managed by onsite–offshore teams acquires a greater significance as it adds to the learnings and knowledge and helps in refining the processes.

CHANGE MANAGEMENT

It is hard to begin any discussion of change management without referencing beaten up clichés like "change is the only constant". Individuals, projects and corporations are subject to constant evolution and change. Change Management has evolved into a significant subject of management and academic review with companies dedicating entire groups of top management to study, plan and prepare for changes in their operating landscape. There is a very strong human angle to change—most individuals are averse to change, unless they can control it.

As the work environments become more global, changes outside an individual's area of specialization, organization and society are likely to have a greater impact on their personal and professional lives. This is especially true of IT outsourcing projects that can easily be impacted by external changes in a local or foreign environment. A manager needs to act as a buffer between the development team and clients, ensuring that they do not succumb to an onslaught of uncontrolled changes in requirements, architectures and designs. The fact remains that not all changes can be forecast or controlled. To address this, managers need to create an adaptable team structure. Some of the key areas of changes that can impact the course of a project include:

- **Changing scope and requirements:** Changes in scope and requirements, unless controlled are among the major causes of failure of projects. Experienced managers learn the art of firmly negotiating the scope and keeping changes in requirements under control. Prioritizing change request and scheduling the requests for a future release are some of the techniques used to mitigate the impact of changes.

- **Code and system changes:** Software application development involves integration of work done by individual programmers. Changes to source code during development, unit and integration testing are a common occurrence. While modular programming ensures that multiple programmers work on the different facets of the same module, integrating their work into a common deliverable is an intricate process. The impact of such changes can be mitigated by use of commonly available tools for software source control and version management. Many of these tools are also being customized to cater to issues of offshore application development. For instance *SourceOffSite*[2] is one of the popular tools catering exclusively to the offsite development needs.

- **Changing business environment:** Business environments constantly change and business leaders and executives revisit plans and strategies to address such changes. Competition, mergers and acquisitions, changes in landscape or business models can trigger a need to change IT systems that support the business. Time-to-market and other reactionary pressures from business may sometimes impact ongoing IT projects.

- **Geopolitical changes:** Geopolitical changes in the business environment can impact the execution of projects. A case in point is the uproar over outsourcing in some communities in the US, UK and elsewhere. There have been instances where managers had to reschedule initiatives bowing to such pressures. Another example is that of a shifting cost base: long term projects that were outsourced by large systems integration vendors for execution at development centers in the US are being revisited in light of the cost arbitrage being provided by offshoring.

- **Changes in regulation and policies:** Governmental regulations, changes in policies and other external changes can adversely affect the performance of projects. For instance, offshore outsourcing projects involve constant international travel by team members. Visa regulations, restrictions or other regulatory changes can adversely affect project plans and schedules.

- **Changes in Staffing and people issues:** Shifts in technology trends and in the job-markets can lead to attrition. Project managers need to plan for attrition and staffing changes during projects. The impact of key members leaving the team can be minimized by constant knowledge sharing and by using tools and techniques of knowledge management.

As professionals become more global, changes outside their organizations and society are likely to have a greater impact on their personal and professional lives. Changes in business, market and environmental conditions at the client site can directly impact the execution of projects offshore. Such changes may sometimes not be very obvious to offshore teams.

BOX 5.1

CASE IN POINT: CHANGE MANAGEMENT

Jai and his offshore team members were working under tight deadlines to develop a digital dashboard for a large financial house. The onsite and offshore team had built a rapport after the key members of the offshore team along with Jai met during requirement gatherings. Besides exchanging regular status reports on project progress, Jai and his peers would also exchange anecdotes and stories on trends and happenings. Jai would also follow emerging news about his client.

During his regular scan of news, Jai came across an article on the impending merger between the client and another

Box 5.1

CASE IN POINT: CONTINUED...

major American bank. Having spent a few years in the US, Jai instinctively realized that such a merger would immediately impact new initiatives including 'non core' IT development as management of the merged organizations began to revisit synergies in their portfolio of applications. During his informal chats with the onsite team members, Jai was able to confirm that although there was no official announcement, his project would definitely go on hold sooner rather than later. He began consulting with his offshore management and stakeholders to formulate a contingency plan.

Sure enough, about three weeks after the merger announcement, the director of the IT division at the client bank announced a major revamp of project schedules including putting certain projects 'on hold' pending further review. Jai and his offshore team was able to effortlessly transition on to a new project as they had already anticipated the strategic shift.

QUALITY

Offshore software vendors have long realized that though cost arbitrage is the biggest seller, assured quality at low cost is what the client really wants. Customers are also generally clear about the fact that though they consider offshore outsourcing to be a strategic choice, the quality of service and operations cannot be compromised. A service delivery Project Manager's view of quality assurance includes focus on internal processes oriented towards ensuring that the application delivers on the promise made to the client. It may be easy to satisfy the stated needs but it is harder for applications to deliver on the implied needs. Just as application requirement gathering exercises could include implied assumptions based on the nature of the client's business or the domain, the quality requirements could also be assumed or implied. In some cases, such implied quality may be obvious.

To address offshoring quality concerns, many offshore vendors take pride in differentiating their quality processes by getting certified by bodies like CMM[3], ISO[4] etc. The process of continuous improvement advocated by the SEI CMM talks about constantly improving on the Key Process Areas: a cluster of related activities that, when performed collectively, achieve a set of goals considered important. The authors[5] [Ethiraj *et al*] in a research paper add that, *"Recognizing the importance of project management capabilities, several Indian firms have been the leaders in adopting CMM guidelines to improve their software development processes.... Meeting these guidelines is not a trivial task. Firms need to make substantial investments in firm infrastructure, systems and human capital."*

The cost of quality, that is, the cost of building quality deliverables may be bundled into the overall application costs. However, specific quality requirements of an application may come at an additional cost because of the additional effort involved in enforcing standards, auditing, verification and walkthroughs translates to added time and effort.

CUSTOMER-VENDOR RELATIONSHIP FOCUS

Customer focus and ensuring customer satisfaction is the prime goal of any IT initiative, more so of offshoring. Customer inputs and interactions impact all the activities performed by the service organizations. This begins with pre-sales phase when a prospective customer scans the landscape and reviews the credentials of the service delivery organization. Project teams interact with the offshore 'customer' that could be the IT team of the client or the onsite team of the same organization. Aspects of interacting and communicating between onsite and offshore teams are also addressed in the communication layer of the OMF. Addressing the risks and perceptions of customers is one of the most import tasks of managers of service delivery organizations. Some of the key areas of customer focus from a global IT team's perspective include:

- **Customer is a partner:** The client may expect the service delivery (offshore) team to proactively address problems and work towards viable solutions. As the offshoring relationship deepens, offshore teams may also begin to assume the role of *technology partners*, proactively suggesting changes and options.
- **Customer may not always be right:** Line of Business managers and the onsite team's technological exposure to cutting-edge trends and industry best-practices may be limited. In such cases, service providers and offshore teams may be expected to have a greater exposure to the trends and bring such best practices from past consulting assignments to the table.
- **Technical and non-technical clients:** The background of clients and onsite teams may not be homogeneous. Some may be very technical and will spell out detailed requirements; others may just have a cursory understanding of technologies and may expect the offshore teams to take a lead. Gauging the strengths and constraints of customers and working closely with business users is an added responsibility of an offshore manager.
- **Customers include internal clients:** For a service delivery teams, customers include internal project sponsors, account managers and all other internal teams that he needs to liaise with. Larger companies generally have centralized support groups that lend assistance on Quality Assurance, infrastructure and procurement of hardware/software. Offshore teams may need to interact with such groups to ensure that the goals of external customers and onsite teams are satisfied.

THE OFFSHORING SWEET SPOT: PROJECT EXECUTION

Finding a common ground, a sweet-spot if you will, between the strategic goals of businesses and innovative applications of technology is an increasing area of focus among IT managers. The goal of technologists is generally focused on providing the best, optimum, cost-effective and contemporary solution to problems of business leaders. In parallel, business leaders continue to focus on extracting the maximum Return on Investment (ROI) on their technology investments. Investments in software applications in large enterprises encompass a vast span of projects that include software architecture, customizing enterprise systems, maintenance and integration of legacy applications; the technology stack may also include niche domains like developing knowledge management and business intelligence systems. Table 5.1 provides a glimpse into some of the key IT domains along with the business decisions they attempt to address.

Conventional view of *IT Strategy* involves mapping business domains into technical focus areas, which in turn is operationalized into projects. Maturity of technology adoption includes formulating technology strategies and proactively helping business leaders identify newer business patterns. Offshoring is one of the enablers of such mapping of business decisions to IT domains. By offshoring the execution of the development cycle, technology managers can focus on bridging the technology-business divide while the offshore partner can work on optimal solutions for technical and operational issues.

Offshoring of IT generally invokes images of low-end system support or maintenance work sourced overseas. The scene is changing with a mix of high-end development and architectural work increasingly moving offshore. Among the wide variety of IT work, the span of technology projects currently being offshored include:

- **Validation services:** Software vendors realize that about 25–40% of the work in software or product development lies in testing and have been increasing their focus on offshoring

Business Decision	IT Domain	Technical Focus
TCO of IT systems: build vs. buy	Software Architecture	J2EE, .Net Linux, Open Source debate, Client Server, migration
Uniform applications for business functions and domains	ERP, Enterprise Systems	Package implementation, Integration
'Business silos.' What to integrate? Why Integrate? How to integrate?	EAI, BPR, BPM	What tools to use? Whether to use COTS, EAI packages, build adapters or web services?
Where is the information we are looking for? Too much data, but very little information	Knowledge Management, Business Intelligence, Content Management	Data Warehousing, Data Mining, Content Management

Table 5.1 Typical Business-IT decisions

validation services. A case in point is Wipro[6], whose VP of Interop solutions was quoted as saying *"We talk to the CTOs (chief technology officers) and CIOs (the chief information officers) of the world."*

- **Re-engineering and maintenance:** One of the core strengths of offshoring is the ability to capitalize on time-zone differences between global locations to provide cost-effective 24x7 application sustenance and enhancements services.
- **Custom application development:** Development of application to predetermined specifications is another core area of focus of offshoring. This includes development in a

wide array of technologies ranging from mainframes to the internet, client-server and, increasingly, even open-source.

- **Enterprise architecture definition:** Architecture definition, generally a high-end activity is done by core enterprise technology teams. The teams may take external assistance from consultants to get focused inputs and for additional research on trends. Research and activities of such architecture definition are increasingly being offshored.
- **System integration:** Integration of Software systems and Hardware components is a big challenge, especially in an ecosystem with ever-changing specifications. The SI offerings that include consulting and solutions to specific problems are increasingly moving offshore as technologies of remote management become more sophisticated.
- **Research & development and innovation:** Project engineering, research and development used to be done in-house. However the trend is shifting as solution exchanges emerge. *"Global solution exchanges throw open a specific scientific and technological problem to the world R&D community, inviting them to solve the same for a specific company according to a pre-defined criteria and in return for substantial rewards and recognition"*, say the authors in a Research-Technology Management Journal[7] article. Offshoring vendors are also increasingly spending on R&D, specifically in establishing technology Centers of Excellence (COE) with software vendors like IBM, Microsoft and Tibco among others.

From an observation of the features of the above list, it can be inferred that although the list is extensive, there is a distinct pattern. Much of the work that is being offshored is well-defined, structured and can be performed with minimal interaction with the end clients. Application developers are as tuned in to the market landscape as IT managers and strategists. The fact that traditional software development paradigms are undergoing a transformation has not gone unnoticed by gurus in the space. In a recent interview,

Marc Andreessen[8], the Father of Mosaic and Netscape browsers, was quoted saying *"Any individual person with a job today that can be done for less overseas obviously is going to feel threatened. But you're not consigning them to a life in poverty. More likely, you're inciting them to get new skills to get a better job."*

The bulk of offshoring IT work continues to focus on aspects pertaining to *'Executing'* (as highlighted in Fig. 5.1) of application development and maintenance, intricacies of which we will examine in the next two chapters. While the execution, context will primarily focus on a phase or a project, a part of a larger offshoring initiative at corporations, managers may occasionally have to deal with small-scale offshoring as the following case 'Small-scale outsourcing to India' highlights.

Box 5.2

ARTICLE: SMALL-SCALE OUTSOURCING TO INDIA[9]

Globalization and outsourcing are turning out to be among the most commonly used words in technology management and executives and managers of all kinds of companies in the US are starting to consider outsourcing. The following is an interesting e-mail from one such executive I received a while ago.

Hi Mohan,

My name is Bob (name changed for this discussion) and I am the CTO of a US-based software company, ABX Corporation. I read your article on the Express IT People website and found it interesting... hence I have determined that you might be an expert or at least know where I might be able to go to get information on my questions.

Box 5.2

ARTICLE: CONTINUED...

For the past two years, we have employed an intern engineer working on his Masters in Computer Science at a US University near our facility. His university visa is expiring at the end of December (after graduating) and he intends to move back to Mumbai. He would gladly accept a job offer from us, but at this point we are not in a position to offer him a full-time position and H1-B status.

We are interested, however, in continuing to work with him from India as an independent contractor with future hopes of starting a software R&D centre in Mumbai. I have a few questions about this:

1. Are there any governmental issues that we (the company) would have to work through, or is he the primary person responsible for setting up his personal status as an independent contractor?
2. I am looking for a report on IT compensation in India. Any ideas on where to go for this?

Please Advise,
Bob
CTO, ABX Corporation

This mail was definitely intriguing and set me thinking. Outsourcing, or at least a flavor of outsourcing, is the kind of remote-work-management being proposed by Bob whose company does not want to throw the baby away with the bathwater, so to say. They want to continue to avail the services of the Indian engineer, who has proved to be good, but cannot remain in the US because of visa hurdles.

Box 5.2

ARTICLE: CONTINUED...

After thinking for a bit, I decided to offer my two cents to Bob, with the disclaimer that this was not legal advice. My reply to Bob went something like this:

Bob,

Thanks for writing to me.

The topic of "outsourcing" (although what you are attempting is not exactly outsourcing) is especially close to my heart. I have researched a number of successful cases of outsourcing for my articles. Several companies ranging from large Fortune 500 giants to smaller organisations are successfully outsourcing work to India. I will attempt to answer your queries as best I can:

Regarding your first question, the candidate could work as an independent contractor billing your company at a mutually agreed-upon rate.

The legal work system in India—although I am no expert, but have a general knowledge of it—is similar to that in the US. The candidate can choose to work for you from his garage (literally). Of course, I am assuming that you will be going through the motions of signing a contract, non-disclosure, non-compete agreement with the candidate before he embarks on the project.

You should also take steps to ensure the protection of your intellectual property rights. In the future, if/when you decide to expand your operations, you will need to form a subsidiary or incorporate a branch in India. (History of EDS's expansion into India makes for interesting reading: http://www.eds.com/india/india_profile_history.shtml).

Box 5.2

ARTICLE: CONTINUED...

The answer to your second question: Most IT outsourcing companies charge clients rates ranging between $12 to $15 per hour for work done offshore.

Although I have not been able to get hold of "official data", the following article should give you some idea: http://www.expresscomputeronline.com/20020415/cover1.s html. Another article on "Current trends in compensation practices in the software industry" might give you some more input on the trends: http://www.mafoi.com/compensation/content/trend7.asp.

It should be noted that your contractor may incur additional expenditure in the form of access to the Internet/intranet, networking, phone calls to/from the US and other logistics that need to be factored in. Hope this helps. Please feel free give me a buzz if you have any other questions/concerns.

Regards,
Mohan

CONCLUSION

The Project Execution layer of the OMF details individual projects and work at the lowest level that is offshored. Technology projects and work of various genera is increasingly entering the realm of offshoring and managers and technologists in both the onsite and offshore sides are beginning to become comfortable with such sourcing. We began the discussion by examining the facets of the Project Life Cycle and the execution of projects conforming to a project plan involving performance of a sequence of tasks and activities. We also looked at aspects pertaining to planning, controlling, change management and a focus on customers and quality. We will continue the discussion on the intricacies of Project Execution, especially on aspects of application development and maintenance in the next two chapters.

NOTES

1. *A Guide to the Project Management Body of Knowledge* (PMBOK® Guide) [Author: Project Management Institute, Inc, 2000]

2. SourceOffSite™ http://www.sourcegear.com/sos/index.asp

3. SEI Capability Maturity Model [CMM® http://www.sei.cmu.edu/cmm/] The Capability Maturity Model for Software (also known as the CMM and SW-CMM) has been a model used by many organizations to identify best practices useful in helping them increase the maturity of their processes.

4. International Organization for Standardization [ISO® http://www.iso.org] is a network of the national standards institutes of some 140 countries, with a central office in Geneva, Switzerland, that co-ordinates the system and publishes the finished standards.

5. Research publication: *A study of firm capabilities and performance in the Software Services Industry* [Sendil K Ethiraj, Prashant Kale, M.S. Krishnan, Jitendra V. Singh]

6. Wipro bullish on validation services [13 September, 2004, SIFY Finance]

7. Global Solutions Exchange: Realizing Extended Innovation in R&D [Deependra Moitra and Mohan Babu K, *Research-Technology Management*: July-August 2004]

8. 'Q&A: Tech innovator looks ahead' Interview with Marc Andreessen [*The Seattle Times*, December 03, 2004]

9. Small-scale outsourcing to India [Originally published as a column in IT People section of *Express Computers*]

CHAPTER 6

Project Execution Layer: Application Development

- Application Development and Software Engineering
- The Development Life Cycle
- Managing the Application Life Cycle
- Offshoring Application Development

The 'execution' part of development involves writing software programs according to design specifications. Offshoring of software application development is a key focus area of project execution. IT leaders are increasingly under pressure to take on proactive roles in addressing business challenges while ensuring that technology is leveraged in a cost effective manner. They are realizing that development of new applications, solutions and products according to specifications using global teams is a faster and more economical way of developing complex applications. Development of application essentially involves teams of architects, designers and developers drawing on several well established principles and techniques of software engineering, working to develop application to address user needs. Offshoring aspects of application development forces architects and teams to bring formalism into their processes.

The fact that traditional software development paradigms are undergoing transformation has not gone unnoticed by technology professionals. Application developers are as tuned to the market landscape as IT managers and strategists. Nowhere is it more evident than in recent job-wanted adverts. Here is a glimpse into two jobs in a recent advertisement from the onsite and offshore ends of the spectrum:

Box 6.1

ADVERT 1: NORTH AMERICA

Wanted: *IT Outsourcing Architect*

Skills: *IT, Architect, change management, management, analysis, telecommunications, networking, desktop, security, firewall, application, architecture, development, disaster recovery, SAS, ISO, Systems.*

Job description: *Interface with other members of the IT Strategic Sourcing team, including potential or existing clients. Provide a holistic view of the scope and costs of moving a potential client to an IT outsourced model. Maintain responsibility for identifying, defining, scoping, and outsourcing requirements related to the responses from RFI's, RFP's, and proposals. Articulate and differentiate financial, technical, and business values achieved by outsourcing major functions or the entire IT function. Demonstrate the ability to understand and process detailed technical discussions, and be able to present that information in simplified business-oriented terms. Understand and identify underlying business assumptions used by a prospective client to ensure they are in-line with outsourcing goals, including ROI, capital investments, service levels, and headcount. Manage organizational changes that are required to function smoothly during the transitional period of an outsourced model.*

It may be observed that though the jobs in Advert 1 and Advert 2 are unique, the offshoring skills are complimentary. The two jobs are from different ends of the onsite and offshore spectrum but they share a common thread—the need for developers to be aware of the intricacies of application development in an offshoring context. These jobs are among an increasing number of those being advertised by employers to address the specific needs of offshoring.

APPLICATION DEVELOPMENT AND SOFTWARE ENGINEERING

Software Engineering, essentially involves the processes of translating business requirements into application systems that satisfy user requirements. There are several software life cycle models described in the Software Engineering literature[1] including the waterfall model, spiral model, prototyping model, the emerging Rapid Application Development (RAD) and Extreme Programming (XP) model and its variants. Managers need to be aware of the engineering aspects of application development along with trends and guidelines from software vendors, industry associations like IEEE[2] and other academic bodies. The Application Development Life Cycle, also known as Software Development Life Cycle (SDLC) includes best practices for gathering requirements, system architectural styles and design patterns to enable development of scalable enterprise-class applications. In addition to the published theoretical fundamentals, application vendors and system integrators also promote best practices and prescriptive guidelines tailored to the use of their tools and technologies.

There is a creative side to computer programming, which no amount of rigor or reference to a stable body of knowledge can replace. It is a well-known fact[3] that the best (top 5% or so) programmers are about ten times better than average programmers. The best programmers tend to be logical and creative and can churn out solid code more efficiently than their peers. There continues to be a debate over how much of software engineering involves pure engineering principles and how much of it is art. Steve McConnell[4] says, *"We still do not have an absolutely stable core body of knowledge, and knowledge related to specific technologies will never be very stable. But we do have a body of knowledge that is stable enough to call software engineering. That core includes practices used in requirements development, functional design, code construction, integration, project estimation, cost–benefit trade-off analysis, and quality assurance of all the rest."*

THE DEVELOPMENT LIFE CYCLE

The focus of the offshoring manager is to enable teams to perform optimally and to facilitate management of processes supporting the development life cycle. A simplistic view of the Software

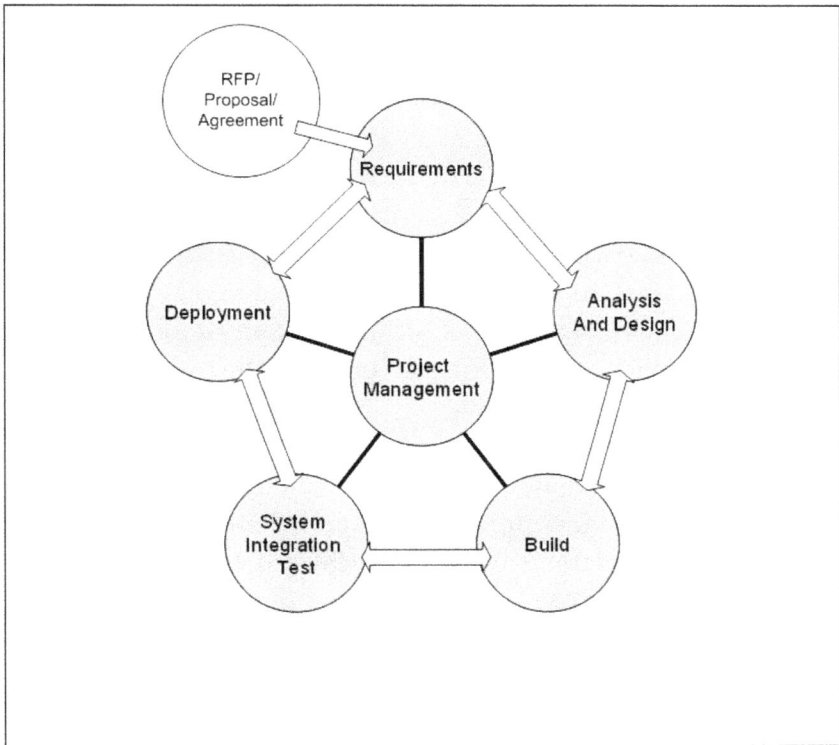

Fig. 6.1   Application Development Life Cycle

Development Life Cycle is depicted in Fig. 6.1. As can be observed, the Project Management block depicted in the center anchors the efforts involved in the different aspects of the life cycle.

Proposals and Agreements

Software Application development projects generally begin with a request from users who give their needs to IT managers to prioritize. Such requests may translate to projects for the development of new systems. Managers may also get a ball-park estimate on effort and cost to get a buy-in and sponsorship from business units. Managers depend on their peers and Subject Matter Experts (SME) and technical architects across the organization to ensure that the proposal is sound and something they are really capable of delivering. The Line of Business managers may either decide to initiate an internal project or work with external vendors. Formal documents calling for proposals include Request for Proposals (RFP) and Request for Information (RFI). In response to such requests, software vendors document proposals that may be a mix of technical specifications, proposed architectures, financials and legal agreements. The process may also involve extensive negotiations on the prices, timelines, effort estimates and other assumptions of the projects.

The proposals and agreements are also an integral part of the contract administration process. A sound contract defining the client and vendor relationship can lead to a structured proposal and guideline for individual units of work and projects. A proposal when accepted and signed off by the client becomes an operating guideline for the project team. Other terms for sign-off common in the services industry include Statement of Work (SOW), Work Order, Work Statement etc. In an offshoring context, work on proposals may be divided onsite and offshore. The onsite team focuses on defining the proposal, whereas the offshore teams may work on defining the proposed solution.

Requirements Analysis (RA)

Requirements Analysis focuses on getting the users and clients to articulate their real need and set the expectations of the proposed system; the focus is on *what to* deliver, rather than aspects of *how to* build a solution. The terminology used to describe this exercise includes Joint Analysis and Design (JAD), brainstorming or Requirements Gathering. Project requirements may be gathered formally or informally. If the requirements are gathered formally from clients, it is preferable that it be signed off to ensure that all the parties are in agreement.

To ensure that all parties are in agreement over requirements, designers or technical architects sometimes build a prototype of the system along with screen mock-ups that everyone can visualize and sign off. When the project and effort estimates or requirements dictates that new and innovative techniques are going to be used, development of demo applications, also called Proof-of-Concept (POC) is also preferable.

The lack of proper requirements gathering can impact the prioritization of effort and the change management activities. Though most of the requirements are gathered during the RA phase, newer requirements, modifications and enhancements may continue to flow-in as the project progresses. Such changing requirements and "scope creep" is one of the major causes of deadline slips and also of the failure of projects themselves. Unmanaged increase to the scope not only impacts the schedules and work but also adversely affects the morale of the people assigned to the project. Experienced managers learn the art of negotiating firmly on requirements.

Managers also need to learn the art of prioritizing the different requirements, especially when there are more requirements than time and resources available. A requirement gathering effort in an offshoring context can be nebulous since the majority of the activities have to be focused onsite. It is preferable that the core design and application team travel onsite during this phase to participate in a thorough brainstorming and value added requirements analysis. Offshoring vendors increasingly provision travel into proposals involving extensive requirement gathering along with development of proof-of-concept. Such strategies may marginally increase the cost of the effort but can minimize risks of communication.

Analysis, Design and Architecture

This is a stage of the project that can set the direction for key technical and architectural decisions. The work breakdown in Architecture definition, Analysis and Design will depend on several factors including the complexity of project, the technologies involved and the business domain. Academics in the Software Engineering community sometimes like to break down the high level planning of software development into discrete tasks involving system architecture, analysis and design; however, there may be sufficient overlaps among tasks to manage these processes as a single cohesive unit. Decisions during this phase may not be purely technical since other factors like the client's preferences and analysis of Total Cost of Ownership (TCO) etc comes into play. Inputs from the client's development and architectural teams may also have to be solicited during such decision making since they will ultimately take ownership of the system.

A major decision a Manager has to take during the Analysis and Design phase of the Life Cycle is whether to document a very high—level design or to work towards a more detailed design. Sometimes the decision may be driven by the clients requirement and preferences but, in many cases, it may be necessitated by who the audience is: if the architects and designers are also going to be involved in building the proposed system, then the team can afford to keep the documentation minimal. On the other hand if the coding team is relatively inexperienced, the manger needs to ensure thorough documentation during this phase. The output of this phase—the architecture and design—acts as a blueprint for the build and unit test.

In an offshoring context, Analysis, Design and Architectural activities will jointly be performed by the onsite and offshore teams. The client facing team will be equipped to review requirements, brainstorm with the client and work with the offshore team on finalizing a technical approach towards the solution.

Build and Unit Test

This stage of SDLC involves translating the design into a workable software code, application modules and programs that addresses the opportunities identified during requirements gathering. During the build phase of a project the Project Manager has to anchor the schedules, tasks and efforts of different team members and ensure that the outputs are synchronized with the project plan. A manager should constantly, albeit unobtrusively, observe the progress and ensure mid-course correction. Managers also need to contend with the team dynamics and be willing to shuffle tasks around the members of the team if there are signs of slack or lack of sufficient progress. During the development stage, the team is responsible for several aspects including:

- **Adherence to development and quality standards:**   This includes ensuring that the code is of sufficiently high quality and adheres to industry best practices.
- **Inter-operable systems:**   The solution development may generally be divided into modules during the architecture and design phase; and such modules are given to individual programmers to code. Ensuring inter-operability of modules coded independently is one of the goals of this phase of the project.
- **Ensuring timely delivery:**   Any delay during this phase of the project can have a cascading effect on the overall budget and timelines. Ensuring delivery on time also means execution of project within budget.
- **Training and mentoring:**   A manager should ensure that there is sufficient mentoring and training of new people in the team. Designers and architects who continue in the project after the requirements analysis and design phase are best suited for such mentoring.

After developing their individual modules, programmers generally test it out in a process called Unit Test. After the modules are unit tested they may need to inter-operate with modules developed by others in the team. The build and unit-test of systems will typically be done offshore, with some interaction with the onsite teams. Communication between onsite and offshore teams including periodic reviews is critical to a successful build.

System Integration Test (SIT)

System Integration testing is sometimes called the *shakeout* phase of the project where the goal is to expose and rectify defects that were not observed when modules were tested in isolation. This may involve integrating the individual modules and programs developed during the build phase into a cohesive unit before testing. This may involve validating the software modules against the design and requirements and checking for interfaces within the application. This will also include testing external interfaces of the programs and modules with other systems in the business environment. Aspects pertaining to security, configuration, reliability etc. may also be verified during this stage of the project. Ensuring that the system can take on the intended load (load testing) and performance is another major key focus area during the SIT phase of the project.

Although the integration test is performed onsite in environments simulating the production environment, it will be jointly co-ordinated by the onsite and offshore teams. System problems, bugs and other issues discovered during such testing may have to be supported by developers based offshore.

Deployment

The last phase of the Life Cycle is the deployment of the new or modified system at the client's and end-user's location. This includes migrating all the code, modules and executables to the 'production' environment. This phase may also include User Accepting Testing (UAT) where end users of the system get together to use the system and sign off on the new functionality if they are satisfied. Such formal testing and analysis of results ensures that the newly developed system confirms to the requirements of the users. Deployment of software systems may also include installation, replication and delivery of code and training the client's support personnel on the operations of the system.

The scope of the deployment phase may sometimes include post-implementation support, also known as warranty support. The warranty phase may involve fixing bugs and supporting application queries during the first few weeks or months after delivery and deployment.

MANAGING THE APPLICATION LIFE CYCLE

Software application development projects are notorious for the high rate of failure. As per a Standish Group[5] survey of projects, only about 28% of the projects succeeded, about 49% were challenged and 23% failed. The report said, *"The reason most of these projects failed was not for lack of money or technology; most failed for lack of skilled project management and executive support."* It went on to add that 97% of successful projects have an experienced project manager at the helm.

Apart from managing the activities of the team on the different aspects of the life-cycle, an application development manager has to focus on the 'big picture' to ensure successful delivery. The discussion on the software development life cycle focused on the processes and the engineering aspects though we did not discuss aspects pertaining to software development of specific technologies. The processes and methodologies are well defined in the general body of knowledge but the actual implementation of the various aspects of the life-cycle during execution of projects may be nebulous, involving a manager's judgment and discretion. Offshoring adds a new dimension to management of application development. Figure 6.2 highlights the Application Development Life Cycle in an offshoring context. It may be observed that the core

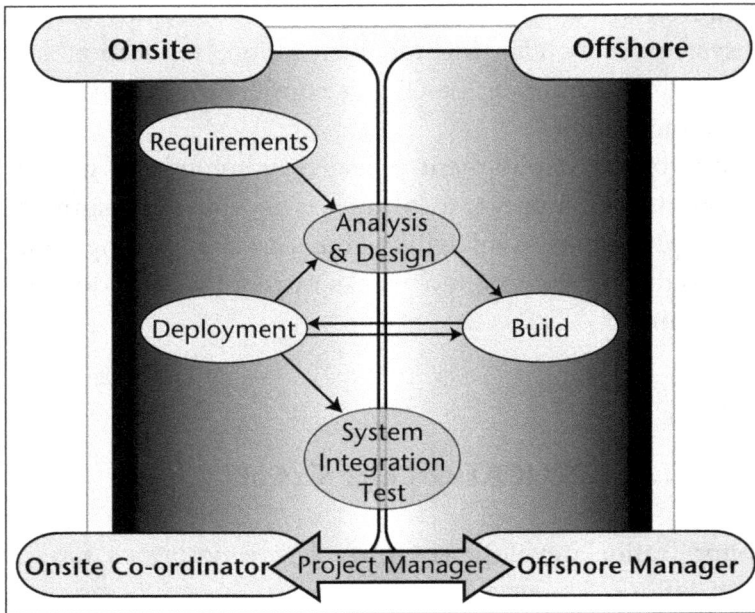

Fig. 6.2 Offshoring Application Development Life Cycle

framework extends from the basic execution context and interactions between project processes highlighted in Fig. 5.1; and in a sense, the offshore-onsite interaction of processes is the backbone for developing an offshoring model for application development.

The key areas of focus of Project Managers during the SDLC phases include adherence to internal and external standards and best practices. Apart from managing the workflow and tasks of individuals of the team and ensuring seamless communication within and outside the team, the execution team needs to focus on:

- **Coding Standards:** The main benefit of enforcing coding standards is to ensure that all the modules developed by the team is consistent and maintainable. Ensuring that the first impression of a code sent from offshore development centers is favorable is one of the benefits of adhering to coding standards.

- **Test Plan:**   Managers generally plan for individual modules and work-tasks to be assigned to developers and programmers. Testing ensures that such code written by developers confirms to the requirements and test-plans that can guide testers are the key to successful testing. In an offshoring scenario, the onsite team may prepare test plans and scripts based on the requirements and then review the test results offshore before accepting the code after development.
- **Integration of Modules:**   Integration of code written by individual developers in a team is one of the prime responsibilities of a manager. After the individuals have finished their build and test, the manager needs to work with module leaders and architects to ensure that all the work tasks integrate into a cohesive unit.
- **System Administration:**   Large development and maintenance projects may have dedicated teams of individuals who manage the various aspects of system administration including version control using tools, regular backup and disaster recovery. Offshore–Onsite teams need to ensure that they allocate dedicated team members to support such system-administration activities for projects executed across the organization for multiple projects and teams to leverage.
- **Configuration Management:** Configuration management, including version control, check-in and check-out of baselined code, taking regular backups are some of the fringe activities that both onsite and offshore teams should focus their attention on. Quality of delivery includes aspects like adherence to agreed upon standards of documentation, ensuring defect-free code and maintainable source-code that adheres to coding-standards. There are several sophisticated tools that can aid in most such activities, and it is the responsibility of the manager to ensure that members of the team use such tools productively.

The key to success of a delivery of software application projects is in understanding the existing architecture of the client, gathering accurate requirements, managing scope-creep and ensuring development on schedule under cost. The development of code during the Execution Layer is perhaps the most crucial aspect of the OMF. At the end of the day, the entire offshoring and sourcing of work to consulting companies hinges on the fact that they employ specialists to work on technical pieces seamlessly. Offshoring of the Application development cycle, if done right, can produce tangible benefits to organizations. A case in point is Viligos which successfully managed the life cycle and released a product in market in record time and under budget by synchronizing the efforts of an onsite–offshore team.

Box 6.3

CASE IN POINT: OFFSHORING APPLICATION DEVELOPMENT AT VIGILOS

Vigilos Inc.[6], a small startup company with a good idea and a strong business model, needs to bring cutting-edge, enterprise software to market quickly on a tight budget. The product needs to integrate disparate physical security systems. The software platform is designed for government and businesses with one or more locations at which intrusion, access control, bar code, point-of-sale, video, audio and other physical security, transaction and operational data is generated.

Solution

The company has two teams, onsite and offshore, working together around the clock to produce one product.

The onsite team of employees is comprised of a few senior level developers and system architects who create the platform architecture, system API, core modules and strategic components. The employees also build use cases and define requirements as well as provide templates and models for the

Box 6.3

CASE IN POINT: CONTINUED...

remaining components. By creating the reference architecture, requirements and user interface models for communicating, the local team focuses the work of the remote to highly leveraged activities.

The remote offshore team builds components based on the use cases and requirements. The offshore team follows the system API and uses core infrastructure modules to fit their tasks into the overall architecture. The Q & A teams onsite and offshore discuss issues using Instant Messenger along with sharing turnover logs and issue tracking, enabling the teams to keep their efforts synchronized. Both teams build, integrate and test their code together daily to ensure a solid fit to requirements, function and design. Defects and source code changes are managed tightly between the two teams. Communications and documentation are standardized and follow defined protocols. Vigilo's ability to cast the work of a few local individuals across the dozens of remote staff in a consistent manner returned predictable results in a very short time.

Results

Vigilos Inc. introduces their product at a high profile industry trade show in less than six months and obtains venture capital funding. Both teams continue to work together and Vigilos' product advances on a compressed schedule and with a budget that could not have supported an onsite team. Customers enjoy the wealth of new product features brought to market by Vigilos.

Follow-up

Over a two-year period the product matured through a number of releases and was installed at several customers. The need to increase stockholder value and their market share drove an effort to increase their sales. To pay for the sales push, Vigilos reduced the offshore team and redirected the funds into their sales team. The local team, with deep product knowledge, remained on staff to provide incremental improvements and support the sales effort. Vigilos remains an innovative company in the physical security market.

OFFSHORING APPLICATION DEVELOPMENT

Managing the application lifecycle and workflow between members of the onsite and offshore teams is perhaps the biggest challenge of offshoring application development. This is a challenge even in application development scenarios not involving offshoring since the task allocation, managing individual units of work aand the traceability between the original requests and end results remains a challenge. The onsite–offshore collaboration and synchronization of communication is a key success factor in offshoring application development. Activities during the different phases of the development could be done onsite or offshore. A brief highlight of some of the typical onsite and offshore tasks in the Execution Layer of OMF include:

1. Onsite Focus

- ***Requirement Analysis:*** Requirement gathering is generally done onsite along with end-users. An onsite co-ordinator may be supplemented by a technical architect and business analyst from the offshore team during requirements gathering.

- **User Acceptance Test:** The audience for this endeavor is the end-user of the application system who begins using it and tests it out. The onsite team co-ordinates the user acceptance test and helps get a final sign-off to ensure successful project closure.

- **Deployment and Handover:** A successful user acceptance generally leads to handover of the system and closure of project. After deployment, there may be a period of warranty support and post-production release issues also.

2. Offshore Focus

- **Build, Development and Unit Test:** The build and unit-test phase is where the majority of effort in an application development project is expended. The responsibility of the build team is to ensure that the code is developed per the requirements and this is an area that requires the maximum tracking and management impetus.

3. Onsite and Offshore Focus

- **Onsite-Offshore co-ordination:** Software service companies that execute service delivery generally designate one or more individuals to take on the responsibility of onsite co-ordination. A similar model can be followed by other teams with onsite–offshore development work. (Ref. Fig. 6.2) The co-ordinators act as the first point of contact for customer interfacing and address technical queries and other system requirement issues. Offshore teams also designate managers or offshore co-ordinators to act as a point-of-contact with the onsite teams.

- **Design:** The offshore team of senior developers and architects get involved in High-level and detailed design of the application and co-ordinates with the client and onsite people. Design is generally an iterative process and may involve several levels of reviews. Onsite travel to participate in joint design sessions may be preferred during this stage to mitigate communication and other challenges.

- **Integration:** System Integration includes shakeout of the system in the client's environment. This is the first time the client's developers and architects get to review the workings of the system after the requirements are gathered and design reviewed. Onsite personnel attempt to ensure that the technical folks at the client's side are satisfied with the performance of the system and that it integrates well with rest of the existing IT infrastructures. The offshore developers may be required to support issues, bugs, fixes and other changes coming out during the integration testing process.

The key differentiator of offshoring, vis-à-vis regular application development, is management of the offshore-onsite aspects of the workflow. The tools and techniques of project management, scheduling, planning and tracking continue to be the building blocks for successful development and delivery; they are extended by the Life Cycle models of the Execution Layer. The breakup of tasks highlighted in the model is indicative based on observations of typical scenarios. Intricacies of specific technologies, the client's comfort with the offshoring model, the robustness and stability of the requirements are among the factors that could determine the actual offshore–onsite mix during the phases of a development life-cycle.

NOTES

1. Popular Software Development Models (Ref: wikipedia.org)

   a. The waterfall model is a software development model first proposed in 1970 by W. W. Royce. Development is seen as flowing steadily through the phases of requirements analysis, design, implementation, testing (validation), integration and maintenance.

   b. The spiral model, defined by Barry Boehm, is a software development model combining elements of both design and prototyping-in-stages, in an effort to combine advantages of top-down and bottom-up concepts.

   c. Rapid application development (RAD), is a software programming technique that allows quick development of software applications. Some RAD implementations include visual tools for development and others generate software frameworks through tools known as 'wizards.'

   d. Extreme Programming (XP) is a method in or approach to software engineering, formulated by Kent Beck, Ward Cunningham, and Ron Jeffries. It is the most popular of several agile software development methodologies.

2. The Institute of Electrical and Electronics Engineers (IEEE) is an international non-profit, professional organization

3. Software Development: Engineering or an Art? [Mohan Babu, *Express Computers'* IT People, 01 Dec' 2003]

4. The Art, Science, and Engineering of Software Development [Steve McConnell, IEEE Software, January/February 1998]

5. *Extreme Chaos* [The Standish Group International Inc., 2001]

6. Vigilos Inc. [http://www.vigilos.com] Vigilos is a provider of Enterprise Security Management (ESM) software based out of Seattle. Case courtesy Paul Thompson of metagyre.com

CHAPTER 7

Project Execution Layer: Maintenance

- 🖳 Application Life Cycle
- 🖳 Maintenance Life Cycle
- 🖳 Challenges of Managing Maintenance
- 🖳 Offshore Management of Application Maintenance
- 🖳 Conclusion

In the previous section of the book, we examined aspects pertaining to management of application development using the OMF. A large percentage of effort of IT professionals is concentrated in sustaining, re-engineering and maintaining existing applications. Burton Swanson[1], in a paper on dimensions of maintenance says, *"The amount of time spent by an organization on software maintenance places a constraint on the effort that may be put into new system development. Further, where programming resources are cut back due to economic pressures, new development is likely to suffer all the more, since first priority must be given to keeping current systems 'up and running'."* This is perhaps one of the main justifications for offshoring application maintenance, re-engineering and enhancement that continue to be the backbone of the offshore development industry, accounting for over 70 to 80% of work done by software service companies. Andre Nadeau[2], executive vice president and chief strategy officer for CGI Group, was quoted as saying, *"If it's maintenance, you can*

155

send 80% to India." Essentially 'India' in the context should be read as any offshored location. By offshoring maintenance, organizations hope to lower total costs while continuing the thrust on new system development.

APPLICATION LIFE CYCLE

The life-cycle of an application starts with the development phase and peaks after it goes to production. After the system goes live, users begin to derive the benefits from the new systems, which could translate to enhancement in their business processes or automating some of the manual work, translating to improved productivity. They continue to derive such benefits till the application becomes obsolete or the business processes change so much that a new IT system is required to support the needs, a point at which the application is phased out. As can be observed from Fig. 7.1, the utility of a software application begins to peak right after development and is consistent during the maintenance phase, till the product is ready to be phased-out after which it declines. Articulating this, Krishnan[3] in a whitepaper adds, *"…the complexity of maintaining the software increases over time because of the increase in the installed customer base and change in the entropy of the system."* During the maintenance phase users may require the system to keep pace with their changing needs. Such changes are one of the main drivers of the application maintenance life cycle. The duration of the maintenance phase of the application could vary based on the business domain, functional area or the nature of the problem. For instance, an application designed to manage the Olympics games may need to be maintained only for the duration of the games, whereas a HR or payroll system may need to be maintained for years.

Application maintenance may need to be performed for various reasons. The initial need for upgrade to application may be to address issues not included in the original developmental scope. This could include usability issues that may only surface after an

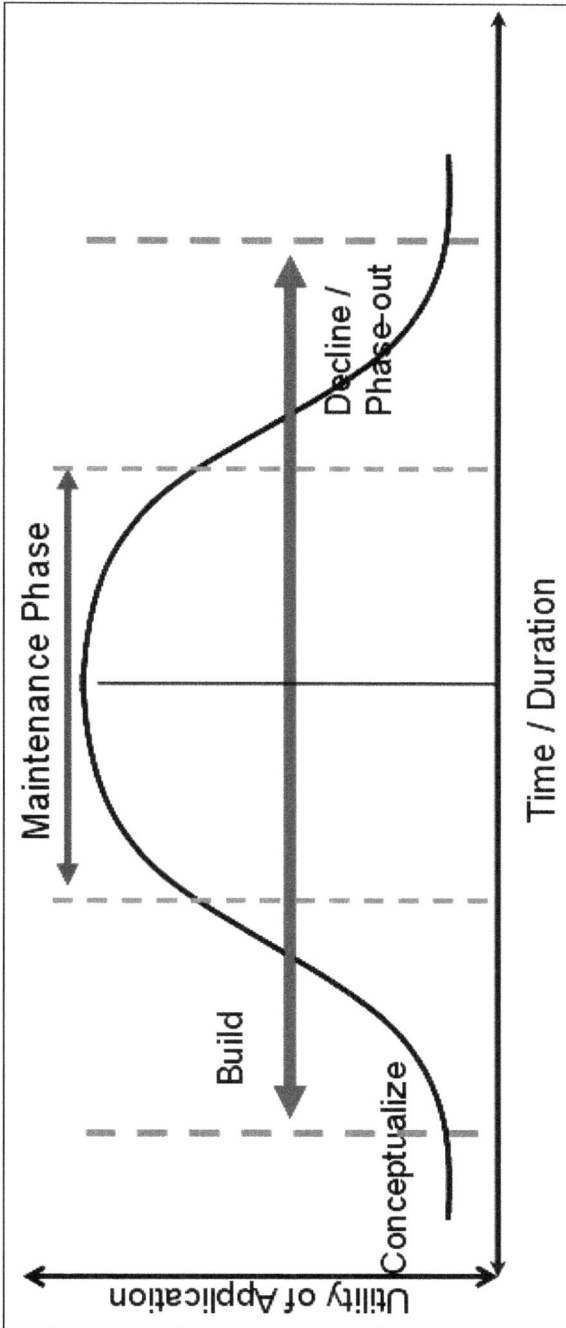

Fig. 7.1 Application Life Cycle

application goes live. After the users begin to exploit the benefits of the system, it may still have to be modified to keep pace with changing needs. Businesses are rarely ever static; changes to underlying business process being supported by the IT application may lead to changing requirements. Catering to the changing business needs is just one of the reasons why IT systems have to be maintained. Major kinds of maintenance include:

- **Production fixes:**   Even the most reliable and fault-tolerant software systems, hardware, machines and networks can fail. Such failures to production systems, also called emergency fixes, may require immediate attention that could include software and hardware updates. These changes are not scheduled in advance but may need to be undertaken on a war-footing. Such production fixes may also be called '24 X 7' support and may be less amenable to offshoring in its entirety, since it may require physical consulting onsite.

- **Application support:**   Users of software applications may have queries on features, workflow and other technical and functional aspects for which a dedicated team may be required. The queries could include aspects of usage of systems, data integrity issues, quick fix and work-around may be supported by having a FAQ and system help documentation. Organizations may have multiple levels of support to address such queries beginning with a help-desk at the first level. Technical queries that cannot be addressed by the help-desk will have to be supported by the second level of support that may be performed by the maintenance team. Such back-office technical support that does not require face-to-face interaction with clients are also increasingly being offshored.

- **Infrastructure induced changes:** Managers of data-centers periodically overhaul the underlying infrastructure to keep pace with the changes in the industry. IT infrastructure changes over a period of time as newer versions of operating systems, system software and upgraded hardware is introduced by vendors. Such changes to hardware, upgrades to system software, operating system, databases etc., may require changes to application systems. Applications may also have to change to keep pace with upgrades to other software and systems that it is interfacing with. Like emergency fixes, these changes are reactionary; however, unlike production fixes, such changes need not be done on a 'war footing' and can be planned and scheduled.

- **Feature enhancements:** Business users and executives constantly endeavor to extract the maximum ROI from their IT investments and regularly request changes tailored to their changing business processes. In the realm of management engineering, this is also sometimes called a *process of continuous improvement*. Changes in software and application standards in the industry may also necessitate changes. Feature enhancements may also involve proactive fixes to systems to optimize efficiencies and to minimize redundancy. Feature enhancement is also called scheduled maintenance and is most suitable for offshoring. Identified fixes or 'change requests' are typically bundled and handed over to offshore teams that make appropriate changes, test them and hand over back to production teams.

Maintenance can involve working on a wide spectrum of technologies, including some arcane and legacy technologies for which only minimal documentation and knowledge may exist. Production fixes and infrastructure induced changes may be reactive whereas feature enhancements could be proactive or planned in advance. Feature enhancement projects that include application re-engineering and fixes are most suitable for offshoring while production fixes may have to be done onsite with some support from the offshore teams. System support and offline handling of customer queries is also increasingly being done by offshore teams. Before we get into details

Box 7.1

Case in Point: Maintenance at a Large Telco

A few years ago, I was associated with an IT team at a Fortune 500 telecommunication company that supported the network engineering and provisioning operations. The core IT application was running a large mainframe database supporting batch and online transactions along with front-end PC based GUI using screen scrapers. It was a complex system designed and developed by a software vendor about 5 years prior to my arrival. About 300 people had been involved in developing the system that took about a year and half to complete. A team of over 85 managers, programmers and business analysts were now involved in supporting the application. This included a team of about 8 people who provided the front end help-desk support to provisioning and network engineers in the field.

The first thing that struck me after joining the team was the fact that there was a lot of domain emphasis which was essential. All the programmers were expected to be familiar with the basics of network engineering and telecom provisioning. Managers would schedule regular visits to a nearby Telecommunication Provisioning Junction to ensure that members of the support team were familiar with the way the systems were actually being used in the field. Interaction with end users also helped team-members build an acute sense of empathy that helped in better problem solving.

A dynamic aspect of our system was the fact that it had to confirm to the Access Services Organization Guide (ASOG) and interchange standards that were defined by members of the telecom industry during periodic meetings. What this meant was that the system had to be upgraded every six months to ensure that it kept pace with the changing standards. There was an elaborate exercise we undertook every six months beginning with the gathering of requirements, basically doing a gap-analysis and mapping the enhancements arising out of the new

Box 7.1

CASE IN POINT: CONTINUED...

standards guidelines. After this, the actual effort would be estimated and a team assembled. The team would work on various aspects of the life-cycle including identifying modifications to the architecture, changes in database structure, changes to screen fields etc. After that the team would work on the design-enhancement-test phases. Multiple teams focused on different sub-domains, worked towards common goal and timelines. As the impact of our changes was on multiple telecom vendors, an extended integration testing phase was needed to ensure that the data files sent back and forth confirmed to the newer standards.

Working on different projects and upgrade releases always involved newer challenges. Working in a stable, yet dynamic environment also helped teams bond, thus creating a distinct culture. The management tried hard to liven the pace by distributing small trinkets and goodies after every successful release.

–The Author

of the maintenance cycle, a case in point from *'Life of a Maintenance Consultant'* based on my experiences from a previous job.

MAINTENANCE LIFE CYCLE

The application maintenance life cycle encompasses infrastructure induced changes and feature enhancements of software applications. Although maintenance life cycle is not a standard term in the industry, we will use it for this discussion as it pertains to the various facets of maintenance of applications. Large data-centers and IT shops devote a good percentage of effort to ensure that the IT systems and infrastructures meet the SLAs stipulated by users. Different Line of Business (LOB) applications will typically have varying SLAs based on their specific needs. An intranet HR portal application, for instance, may not have the same requirement of fault-tolerance as a customer relation management database. If the CRM system is down when a customer calls, he could walk away to a competitor; similarly, the consequences of a stock broker's trading system going down during peak-time could lead to huge losses; hence the designers and managers will build multiple levels of fault-tolerance and redundancy into such systems.

Data-centers and IT shops devote a good percentage of effort to ensure that the IT systems and infrastructures meet the SLA stipulated by users. Figure 7.2 depicts a typical maintenance Life Cycle. The first level of support for LOB application is handled by tech call-centers, and in their absence, by the production support staff. The call center or help-desk may be supplemented by 24 X 7 production support and other technical personnel who attempt to address the IT and application related issues faced by users of the different systems. The issues could pertain to system faults, hardware errors, network problems or could stem from the lack of awareness of the features or other usability related issues. The call center personnel need to be aware of the features of the systems and should also have an understanding of the underlying technical architecture as it pertains to the business workflow. They may refer to internal FAQ databases, system references, documentation and other support repositories while attempting to address the problems of LOB users. Such databases may already have records of known 'bugs' and system problems along with a description of work-arounds and solutions. Problems that are unique or encountered for the first time may have to be addressed and recorded in the change management systems.

IT shops and data centers that maintain applications have systematic management practices to track the workflow of changes beginning with the original request from business users. Such

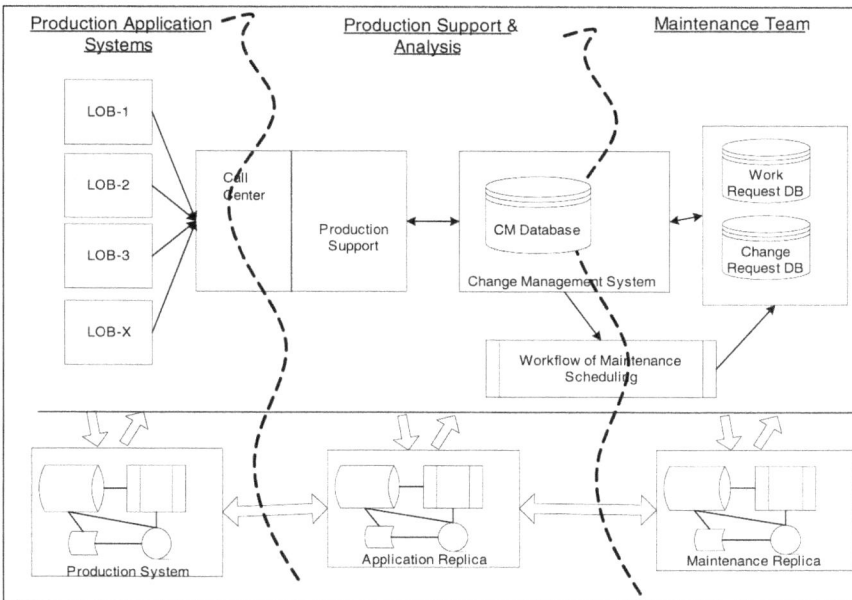

Fig. 7.2 Maintenance Life Cycle

requests may have to be tracked all the way through internal scheduling and decision making processes that lead to scheduling the changes. The processes also ensure that the changes are incorporated as agreed. Such a traceability from requirements to production is essential to ensure customer and user satisfaction. This change management process also feeds into root-cause analysis to examine the major points of failure of systems. Maintenance of applications requires systematic management practices to track the workflow of changes beginning with the original request from business users.

Technical challenges in maintenance include identifying the root cause. Lack of system documentation and use of outdated techniques and tools may make the process of debugging harder; similarly coding and testing of applications may require upgrades to systems developed using legacy programs, skills which may be hard to find in the marketplace. A strong need for domain skills is another requirement for teams maintaining systems since the end-users are going to be business savvy and the teams may not always have the luxury of business-analysts or systems-analysts to interface. Another key technical challenge is the need to replicate the entire system from production to application support and maintenance environments. Offshoring of system maintenance implies the need to maintain a replica of the system offshore. The replica should be refreshed periodically since the system may not be synchronized with periodical updates.

In Figure 7.2, we highlighted the workflow of maintenance scheduling that involves a series of activities. Business analysts, representatives from LOB and IT management may periodically get together to do a trend analysis of the commonly occurring problems in different LOB applications. Figure 7.3 depicts the typical workflow of scheduling tasks for maintenance. During the root-cause analysis, the team may also come across opportunities for improvement of both business processes and the underlying IT systems supporting them. The trends analysis also includes analysis of the cost-benefits of the problems and identifying patterns in the occurrence of problems. Changes to business process may induce changes in the usage of system, and expose flaws in integrating the industry best practices or user training that will have to be addressed by the LOB and business teams. Changes to IT systems including system fixes, software enhancements and adding new functionality may be identified during this phase. If the problems are frequent and fixing them may be cost-effective, they may be scheduled either as change-requests or work-requests. Collections of such tasks are turned over to the maintenance project manager who manages the workflow of tracking the solutions till they go to production.

Tracking issues and fixes in a maintenance cycle may include usage of terms like *issue tracker, change request, trouble ticket, bug request* etc. The actual process of tracking and ensuring that changes are made and prioritizing the changes may be done by a formal management group called the Change Control Board (CCB). The processes also ensure that the changes agreed-upon are incorporated. Such a traceability from requirements to production is essential to ensure

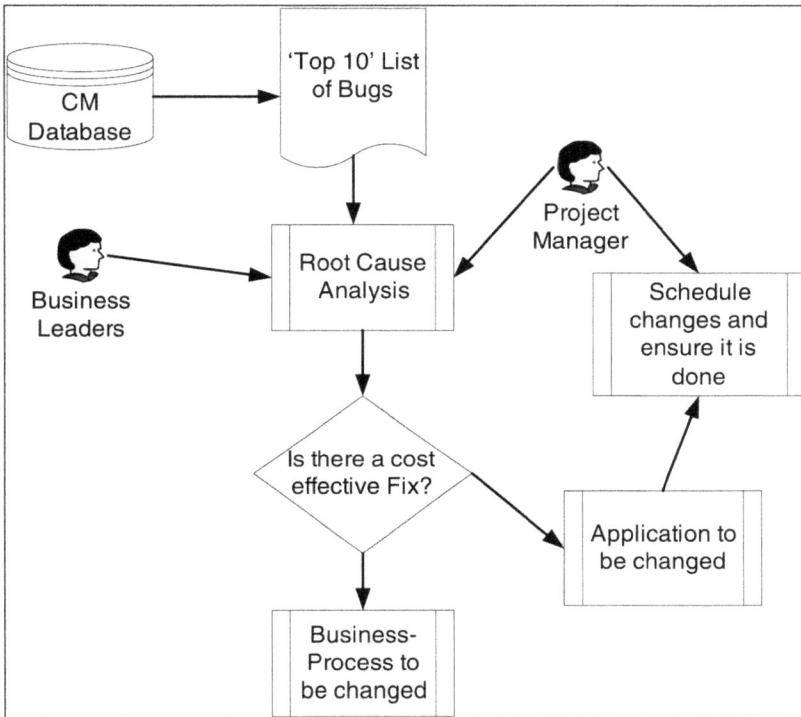

Fig. 7.3 Workflow of maintenance scheduling

customer and user satisfaction. This change management process also feeds into root-cause analysis to examine the major failure-points of systems.

OFFSHORE MANAGEMENT OF APPLICATION MAINTENANCE

IT managers have the dual challenges of addressing the 'innovative vs sustain' dilemma; they have to focus on newer development to address and capitalize on technical challenges while at the same time ensuring that existing systems continue to meet the demands of users. There is, therefore, a very strong business driver for offshoring: by sourcing routine maintenance of applications, the development teams of client can get more involved in development, R&D and core engineering activities. The Global Delivery Framework that we examined in the previous chapter can be extended to manage application maintenance initiatives also. Planned maintenance and working towards fixing system bugs according to predefined SLAs provides a structured, measurable means of managing offshoring. Figure 7.4 depicts the maintenance extension of the Execution Layer of OMF.

The workflow is managed by the production support team onsite and by the maintenance team offshore. Offshore project managers work with the onsite project production support teams to prioritize the schedule of the tasks in change-requests. They also work with the Customer Support teams to replicate the problems and validate the work being done. Change requests are merged

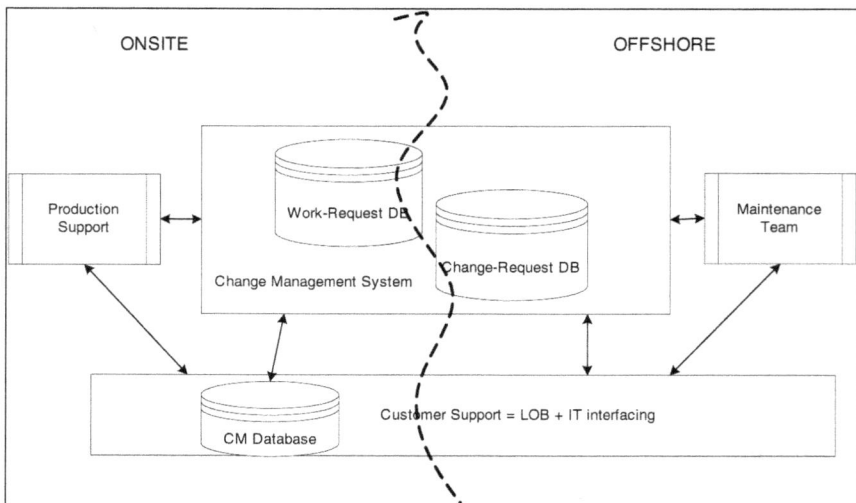

Fig. 7.4   Maintenance extension to OMF

together and addressed in scheduled project releases. The offshore team may manage development on a replica of the production system after which the changed code is sent to the onsite team that validates it before migrating it for the integration test.

The fact that IT departments typically schedule maintenance releases periodically, which could be monthly, quarterly or annually, may help in scheduling offshoring in a structured manner. Offshore teams have clear deliverables and outputs to work toward and can ensure support consistent with the expectations. Project management of such maintenance releases involves managing the workflow beginning with prioritizing and scheduling individual tasks identified. IT shops typically have multiple projects catering to the needs of individual LOB applications scheduled for release together. Such large maintenance projects need to be co-ordinated by dedicated program managers to ensure that all the core system functionality including production migration, management of user training, system down-time etc are synchronized. This also means that managers of the different application projects being maintained work towards commonly agreed schedules. A case in point is a successful offshoring initiative from SunGard (Box 7.2).

Box 7.2

CASE IN POINT: SUNGARD'S MAN IN INDIA[4]

Mack Gill hopes to expand the financial service giant's offshore services outside the company's walls.

Catching up with the president of SunGard Offshore Services (SOS) these days will probably involve voice mails and e-mails, as Mack Gill splits his time between his firm's Manhattan and its Bangalore or Pune offices in India. Gill is

Box 7.2

CASE IN POINT: CONTINUED...

currently in the midst of expanding the SunGard business unit's customer base from strictly internal SunGard clients to include the financial service firm's external client base.

"SOS was a well-kept secret within SunGard," says Gill. The Indian unit was formed in 1993 to act as an internal development team for the company's Brass trading platform. "The team grew nicely through the Nineties and in a couple of years it grew quite sizable. It has become a resource used by many of the other SunGard units," he says.

Although his organization shares the same initials as the international Morse Code distress signal, Gill doesn't think this will be a hindrance as SunGard re-brands itself and spins off its business continuity arm, which will keep the SunGard moniker. "To be honest, the initials aren't going to last with the re-branding," says Gill, whose business will be part of the newly branded company.

Since other brand-name offshore service providers, such as IBM, Tata Consultancy and Wipro, also provide project development, data migration, system re-design and quality assurance services, Gill differentiates SOS's offering by its depth of domain knowledge. "If you look at the offshoring market, domain expertise is hard to find," he says. "It's domain expertise and the focus is on smaller scale deployments, such as department-level contracts." At this point, all of SOS's customers are other SunGard business units, but Gill plans to change that by offering three key features to investment firms—quality, economics and "expertise around SunGard products," he says

What has Gill's mandate been since taking the helm of SOS? The answer is simple, he says. "As any other SunGard

BOX 7.2

CASE IN POINT: CONTINUED...

unit president, my mandate is to grow the business and find interesting ways to do that." Unlike the pre-spinoff SunGard, Gill has set his sights on building his organization through aggressive organic growth rather than through acquisition.

Since the beginning of the fourth quarter last year, SOS has grown from 200 developers located in Pune that have been supporting 13 other SunGard business units to more than 250 developers. Gill expects far larger growth throughout the year, but he declines to offer any details. "There are too many growth opportunities out there," he says. The major factor contributing to this rapid growth is SOS's presence in Bangalore and Pune. "Bangalore is the Silicon Valley of India, and as such, it enables you to source certain types of talent. Pune is the biggest education center with a focus on engineering in India. It's the Oxford of the East," he explains.

The opening of the unit's Bangalore office has only happened in the past few months, as SOS chose to leverage sibling business unit SunGard SCT's presence within the city. "SunGard SCT serves a different part of the market," explains Gill. "Historically, it focuses on the higher education market by providing enterprise resource planning systems for universities in Europe and North America. That unit will continue to focus on higher education, but we will use the facilities and management. Now we can scale twice as fast by being in both cities," he says.

Gill is also looking to grow SOS beyond its current Indian borders. "We're looking to widen the area and are looking at other geographies," he says. "SunGard already has development locations in Israel, New Zealand, Canada and other areas."

Box 7.2

CASE IN POINT: CONTINUED...

Resume in a box

Name: John Mackay "Mack" Gill

Education: The University of British Columbia and Yale University

First Job: Working at the United Nations' Department of Political Affairs in New York, helping manage negotiations during the 1993 Haitian crisis.

Best Advice: "Happy clients buy more software."

Window or Aisle?: I like a window seat—you don't want to miss the Himalayas en route to Mumbai.

CHALLENGES OF MANAGING MAINTENANCE

Offshoring of maintenance ensures that the intricacies of managing changes, unit testing and adherence to internal guidelines are transparent to the onsite team. Maintenance projects are distinct from the new application development initiatives since the process begins with a suitable code-base that the output will also add to. In Fig. 7.2, we saw the activity called "Schedule changes and ensure that it is done". An application maintenance manager is responsible for orchestrating the work involved in changes. Some of the major responsibilities of maintenance include:

- **Replicating user problems:** The maintenance teams receive change requests from a bug database or other lists. Problems listed by users need to be replicated in a non-production environment and analyzed before a solution can be designed. For this, maintenance teams manage replica of the production systems that they use for testing bug fixes, reproducing problems etc. Replicating the user problems in isolation while being hundreds of miles away can be challenging, especially since the sequence of activities performed before encountering the problem is not very obvious. Project managers may need to liaise with users and managers of LOB systems if they need assistance in replicating the user problems since they may involve a set of complex user tasks. Offshore teams may not have a replica of production data available to them due to security and other restrictions of moving customer data offshore. The *dummy* data may not always suffice in debugging complex problems and the offshore team may have to invest in other sophisticated tools to 'view' onsite systems at runtime. An example of such a software is the Client for Microsoft Network[5].

- **Customer interfacing:** A maintenance manager has to actively participate in discussions with business users and others to ensure that the system is being maintained as agreed. This also helps managers get a 'big picture' of the issues that can help in prioritizing tasks according to the business severity. Customer liaising is also essential to define and ensure satisfaction with the SLA. In an offshoring scenario, onsite co-ordinators will have to take on the responsibility of client interactions and abstract the dialogue of such interactions from the offshore teams to make them more productive.

- **Scheduling releases:** Maintenance and changes to production systems are rarely done in an ad-hoc manner. Releases are scheduled periodically and fixes included logically according to business needs. In large organizations, many LOB applications may be maintained by multiple IT groups; each LOB application in turn may be managed by several teams and managers. All the teams have to synchronize the development and enhancement efforts in order to provide a seamless user satisfaction. Offshore teams need to be aware of the maintenance calendars and work towards such periodic releases in conjunction with the IT managers liaising with LOB owners.

- **Task allocation:** Maintenance of business applications can span months and years. A manager has to ensure that the tasks are allocated according to the capabilities of the team and to ensure that all members of the team are sufficiently motivated. This is a challenge common to regular maintenance and offshoring scenarios. Participating in maintenance teams is sometimes perceived to be non-core engineering activity that is not 'glamorous.' The authors in an ACM whitepaper[6] add, *"The maintenance and enhancement of operational aspects of software systems is frequently viewed as a phase of lesser importance than the design and development phases of the system life cycle."* Managers need to creatively motivate individuals to get interested in such maintenance and support roles too.

- **Configuration management:** Configuration and change management is the most significant activity managed by project managers of maintenance initiatives. Changes of configuration can be accentuated by offshoring. Best practices attempt to ensure that the changes requested by managers are incorporated into the system without any major disruption to the existing functionality. This also means that the manager needs to control the traceability between the original change request and the actual fix that is incorporated into the system. The change to the code is generally done in the test environment by developers. Such test environments may be a replica of the production system but may still not have all the live interfaces; hence issues of incompatibility may surface after the changed subsystems go to production, which the manager needs to guard against. Several automated tools of configuration management[7] exist, some explicitly tailored to address offshoring challenges.

CONCLUSION

IT managers and business leaders are already beginning to take stock of the benefits of offshore outsourcing. Maintenance of legacy systems are emerging as the most popular candidates for sourcing because of several reasons that we looked at in this section. We also examined how the framework for maintenance of the Project Execution Layer can be extended to manage offshoring projects and initiatives too. In our discussion, the practices and activities of offshore maintenance have been examined in a technology and vertical agnostic manner. Specific business needs, technologies and vertical dimensions will have to be considered while formulating an offshoring strategy for application maintenance.

NOTES

1. The Dimensions of Maintenance [International Conference on Software Engineering, E. Burton Swanson]

2. CIOs on Offshoring [Allan Hoffman, Monster Tech Jobs Expert, http://technology.monster.com]

3. *Software Release Management: A Business Perspective* [Mayuram S Krishnan, IBM Centre for Advanced Studies Conference]

4. SunGard's Man in India [Rob Daly , Waters, 02 February 2005 http://db.riskwaters.com/public/showPage.html?page=205238]

5. Microsoft®: Client for Microsoft Networks The Client for Microsoft Networks component allows a computer to access resources on a Microsoft network.

6. Characteristics of application software maintenance. *Communications of the ACM Volume 21, Issue 6 (June 1978) B. P. Lientz, E. B. Swanson, G. E. Tompkins*

7. SourceOffSite™ http://www.sourcegear.com/sos/index.asp

CHAPTER 8

Communication Layer

- Communication Context
- The Communication Layer
- Tools and Technologies of Communication
- The Communication Layer
- Conclusion

Groups and individuals are increasingly finding themselves in global and multicultural teams as offshoring takes off. Managing such globally distributed teams and onsite-offshore co-ordination requires clear, unambiguous communication between individuals and teams, and communication in a business context continues to be a significant area of focus and concern. Not surprisingly, Peter Drucker, the renowned management guru was quoted as claiming[1], *"that 60% of all management problems are a result of faulty communication."* The problem of poor or ineffective communication gets accentuated when one begins to look at workings of projects spanning geographic and cultural boundaries where the chain is only as strong as the weakest link. The PMBOK[2] addresses communication management issues adding that rigorous planning, *"provides the critical links among people, ideas, and information that are necessary for success. Everyone involved in the project must be prepared to send and receive communications, and must understand how the communications in which they are involved as individuals affect the project as a whole."*

COMMUNICATION CONTEXT

The basics, including the theoretical foundation of communications management continue to form the underpinning of onsite and offshore co-ordination. The PMBOK[2] focuses on four major aspects including communications planning, information distribution, performance reporting and administrative closure. Planning includes determining the communication needs of the stakeholders; basically on aspects of who needs the information and how to get it to them. Information distribution focuses on actually getting information to the stakeholders in a timely manner. Performance reporting includes collecting and discriminating information and reporting. The following are some of the theoretical underpinnings of communication across cultures that may come into play in an offshoring scenario:

- Hall[3] talks extensively of 'High Context' and 'Low Context' cultures. High-Context cultures are relationship centered while Low-Context cultures are task centered.
- In his writings, Hofstede[4] deals extensively with aspects of cross-cultural management showing a high variability between countries that include *individualism/collectivism, power distance, masculinity/femininity, uncertainty avoidance.*
- Another dimension of discussion on cultures involves direct/indirect cultures. *Direct* cultures meet conflict head-on whereas *Indirect* cultures avoid conflicts; similarly *Expressive* cultures display emotions openly while *Instrumental* cultures keep them hidden.
- Cultures are also categorized as formal/informal. Formal cultures value formal protocols and customs while *informal* cultures don't value such protocols as much.

During interactions with international colleagues, peers and clients, emphasis on both formal and informal communication techniques are of equal importance. In some cultures, the formal and informal communication may be de-linked: project meetings will only comprise of work related discussions, with individuals managing all personal communication during lunch or at coffee breaks. Cultures may also be highly hierarchical where the project manager is looked-up upon as a de facto leader and an arbitrator of discussions. In other cultures, especially in the western culture, informal and formal small-talk is the way team members gauge each other. For instance, talking about one's football or soccer team, may be a way of bonding with co-workers. The cultural aspects of communication are increasingly being capitalized by IT managers. Case in point: the authors[5] give the example of, *"people originally from India, but with higher education and long-term residence in North America, have been reposted to India as expatriate managers for outsourcing projects. Such managers are often effective in overseeing complex outsourcing projects."*

Aspects pertaining to soft skills, like body language, use of phrases and expressions also come into play while dealing with people from other cultures. The interaction between Raj and Carol in Fig. 8.1 describe one such example of miscommunication while using idioms and phrases that have a certain meaning in one culture but can convey a converse meaning in another culture. Indian teenagers regularly use the expression freak-out to imply that they enjoyed something thoroughly; in the US *'freak out'* means spooked! There are several online resources[6] that have listings of regular expressions and idioms in different parts of the world.

A manager needs to ensure that the teams located onsite and offshore need to work as a cohesive unit. The risk of offshoring communication can be mitigated by conscious effort and by deploying some of the tools and emerging technologies, including blogs that simulate a water-cooler conversation, the occasional video-conference to put a *name-to-the-face* and travel between onsite and offshore if feasible. The focus of offshoring communication begins with stakeholders, especially with onsite teams communicating the initiative clearly to the rank-and-file (Ref: Box 8.1). Such communication sets the ground for a buy-in from those at the project execution end.

Fig. 8.1   Expressions and phrases in cultural communication

BOX 8.1

CASE IN POINT: COMMUNICATING OFFSHORING TO EMPLOYEES

If you work for a company that is contemplating sending technology work to India or another country (or has already), you will be well-advised to keep these 10 simple rules handy:

1. Your employees are the most important audience you have. Treat them with respect and dignity, even if you're telling them things they don't want to hear.
2. Make sure your communications plan is intertwined with your operational plan. One cannot succeed without the other.
3. In order to achieve 2, make sure you have a communicator on your operations team from the very beginning of the

Box 8.1

CASE IN POINT: CONTINUED...

process. He or she will ask questions along the way that you'll get from your constituents.

4. Realize that different audiences will react differently to the news that you're engaging with global partners. Some will be supportive, and others will be anything but. You should develop messages that are consistent with one another, but that appeal to each audience.

5. Even if you are absolutely certain your offshore program will never result in layoffs, do not make that promise. You can't always predict the future, and it's very difficult—not to mention ill-advised—to back away from a promise once it's been made.

6. Prepare to be put on the spot about your offshore program. There will always be people who are opposed to your strategy and will want to take you to task for it. For that reason, you should expect the ambush and be prepared when it happens, rather than whistling past the graveyard hoping it doesn't.

7. Go proactive when the time is right. Sometimes the best defense is a good offense. Make sure you have all the questions answered before they're asked, and then explain your reasons for going offshore without apologizing for it.

8. Convert the "undecideds" in the middle. An equal number of employees will support your program as oppose it. But a far greater number will be undecided until you communicate with them. So focus your efforts on these undecideds. Communicate candidly, openly and often with them about your outsourcing program. Delete the hype and don't duck the hard questions.

9. Place greater communications emphasis on *why* you're going offshore, as opposed to *how* you're doing it. If people understand the rationale behind your decisions, they're more likely to support the actions you're taking.

Box 8.1

CASE IN POINT: CONTINUED...

10. Stake out a position on outsourcing and stick to it. If you're going global, explain it and treat even the naysayers with respect. If you've decided against going global, say that too. Whatever your decision, don't waver.

–Tom Phillips[17]

THE COMMUNICATION LAYER

The Communication Layer of the Offshoring Management Framework cuts horizontally across all other *Layers* of offshoring. The implication is very clear: successful foundations of communication management are key to the success of offshoring. Cultural aspects of communication like challenges of communicating across different languages, dialects and accents along with other cultural traits also come into play in offshoring. The communication layer builds on the fundamentals of business and technical communication management principles. As highlighted in Fig. 8.2, the dynamics of interaction across and between the layers have two distinct pairs of dimensions— the offshore and onsite, time and space—encompassing the interaction between the other three layers.

The focus of communication planning will be on the four dimensions of time, space, onsite and offshore and will continue to build on the fundamentals of communications management. The yin-yang between onsite and offshore interaction and co-ordination is the highlight of globalized management and manifests itself in all the layers. As a part of their governance activities and planning, senior executives and stakeholders set the context for offshore–onsite

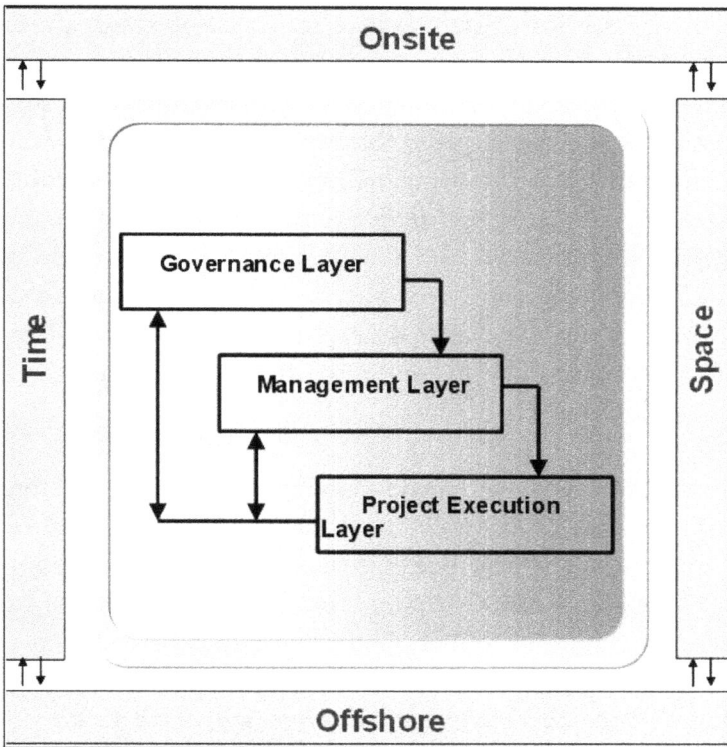

Fig. 8.2 Communication Layer

communication. The activities in the management layer ensure that the projects and programs operate according to strategic direction and plan for communication including use of modes, tools and technologies. The Project Execution layer implements the onsite–offshore communication at the lowest level of granularity. The time and space dimensions highlight aspects of time-zone differences and boundaries between geographic, cultural and organizational silos. The physical distance between teams along with time zone differences can either be an advantage, as in case of teams working towards 24 X 7 support, or a disadvantage, as in case of teams spread across geographies trying to collaborate.

Organizational policies and practices from both onsite and offshore teams will guide the communication management planning endeavor. Such planning will ensure that all the stakeholders know what to expect in terms of communication and what their responsibilities would be. This includes definition of the information distribution mechanism: defining the modes of *who* needs the information, *what* they need to know, *when* they need the information and *how* they need it. The constituents of planning may include:

- **Stakeholder Analysis:** The stakeholders at the different layers of the OMF will have varying communication requirements. For instance, the stakeholders at the governance layer, the senior executives are going to focus on processes, while the stakeholders of the management layer are actually going to define the processes and the stakeholders at the project execution layer will have to ensure that the processes are followed.
- **Communication Management Plan:** The published plan will articulate the workings of communication flow along with addressing the needs of stakeholders. The communication management plan will be tightly integrated with other project planning exercises. The plan will articulate the distribution structure, the description of information that will be distributed, the modes of communication and the tools and technologies that may be adopted.
- **Communication Templates:** Planning will include defining templates to facilitate communication; for instance, templates for recording the minutes of meeting, status reports, requirements gathering and other routine activities will help parties communicate effortlessly.
- **Scheduling Publication:** Information for stakeholders may have to be scheduled differently based on their varying

needs. A calendar for publishing information will help manage the expectations of stakeholders.

- **Change Management:** Even the best laid plans may be subject to change; therefore a communication management plan should consider changes in mode, frequency and technologies adopted for communicating.
- **Information Retrieval Mechanisms:** Information that is already published may need to be retrieved by stakeholders and others at any point in time. Communication management needs to consider such information access and retrieval and achieving needs.

The focus of discussion in the communication layer shall be on the tools and technologies that play a vital role in facilitating a smooth interchange in bridging the dimensions of time and space, with onsite and offshore.

TOOLS AND TECHNOLOGIES OF COMMUNICATION

There are several sophisticated collaborative tools and technologies available in the market. Such tools facilitate the management of workflow and activities during the different stages of offshoring. The communication and collaboration technologies fall into two broad categories: Asynchronous (or offline) communication and Synchronous (or online) communication, both of which have their merits. Offline tools of communication like blogs, e-mails and other broadcast mechanisms are suitable to bridge the time-zone gap while online communication includes VoIP phones, video and audio conferences and instant messengers that provide real-time communication, bridging the space and distance factors. Managers also use project management and workflow tools to share information, status and updates across teams. Some of the key trends include the increase in adoption of internet and network based technologies connecting people across geographies.

The technology drivers include:

a. Business correspondence over e-mail has increased
b. Global teams are requiring seamless access to information through shared databases and common repositories
c. Web pages, wikis, blogs and other tools of collaboration are promising enhanced user experiences
d. Continuing focus on communication by voice and video; and the re-emergence of videoconferencing to bridge the space gap and mitigation of travel
f. Emergence of collaborative application suites including multi-point data conferencing, sharing documents and real-time digital dashboards
g. Growing acceptance of project and workflow management tools and systems

We will highlight some of the tools and techniques used to facilitate communication across geographies that include e-mail, instant messengers and web-logs (blogs). Although there are more basic communication tools and channels including telephones, face-to-face communication etc., we will not focus on the intricacies of such modes of communication since there is sufficient body of knowledge in the management literature; instead, we will focus on a few emerging and nascent tools that are facilitating globalized teams.

E-mails

Stating that the ubiquitous and pervasive access to e-mail has changed the way we do business in the digital age would be clichéd, but apt. The pervasiveness of e-mail technologies has made global communication effortless and inexpensive. For most of us, e-mails have almost replaced office memos, newsletters and other forms of paper communications that were routine even a few years ago.

E-mails have also become the preferred mode of communicating between onsite and offshore teams in a global context and are regularly used between clients and vendors.

E-mail communication in a globalized project context takes on additional dimensions of cross-cultural communications. Offline e-mails supplemented by online communication such as voice or instant messages are a good way to avoid ambiguity. The caveat is that cultural aspects such as written idioms and colloquialisms may creep into messages that may be out of context in certain cultures. E-mails can serve as an excellent tool to share and record minutes of meetings, broadcast messages to groups of people and point-to-point communication between individuals. Other advantage of e-mail include the ability to send different formats of attachments to supplement messages that could range from documents, video and voice files, diagrams and links to relevant websites.

Use of e-mails in an offshoring context may require additional focus and strengthened policies and practices that both onsite and offshore teams may need to jointly formulate and regulate. Some of the key aspects of focus include:

- *Project e-mail policy:* Both the offshore organization and vendor may have organizational communication policies. Offshoring projects and engagement managers may also formulate policies for interchange of communication and e-mails to streamline the process and avoid an *information overload*. Understanding and articulating the policies in place are a part of the project planning process.
- *Confidentiality and secrecy:* Managers need to sensitize teams on the significance of e-mail confidentiality since corporate information can easily be passed out via e-mail. In an age where intellectual property is highly valued, such a leak can be disastrous.
- *Security and virus hazards:* E-mails, especially with macros and attachments constitute a big security risk. Security policies should include the use of tools to detect

viruses and malicious software including Trojan horses, viruses and worms.

- **Legal issues:** Legal issues which can arise from the use of a corporate e-mail account can include copyright, intellectual property issues and contractual obligations among others. For instance, an e-mail sent using a company's e-mail account could be construed to be a legally binding document, leading to cases where the company could be obligated to enter into a contract unwittingly. Vendor and client teams may need to be sensitized in the offshoring context.

- **Expensive network usage:** Non-business related e-mails are increasing at a fast pace. Such e-mails (with attachments) could clog the corporate network, leading to unnecessary bottlenecks. Spams are another common scourge of networks. To mitigate such risks, teams should be discouraged from the use of official mail IDs for personal registrations and other uses on the web.

- **Employee relations issues:** Sending e-mail messages containing material that may be construed as being racially or sexually offensive could be especially tenacious. (Ref: Box 8.2). Sensitivity to cultures, mores and norms of the societies where teams are based need to be factored in while formulating e-mail usage policies.

Large organizations may have formal policies governing the use of e-mails, networks and internet tools. The use of e-mails across organizations is also subject to general practices and mores. As the economies across the globe get more integrated, Project Managers supervising projects spanning international boundaries need to become aware of the restrictions on e-mail and international communications. E-mailed documents are increasingly acquiring legal status[7], and contracts and SLAs exchanged by e-mail are increasingly being viewed as contractually enforceable documents. In global organizations, especially in organizations where projects are outsourced, security is a key driver and may face additional scrutiny. E-mails and corporate communications came under the spotlight with the spate of (in)famous bankruptcies and ensuing investigations. In a number of prominent cases, e-mails sent by ex-employees have acted as crucial pieces of evidence used by prosecutors. In certain industry verticals, for instance in the financial sector, governmental regulation—like Securities and Exchange Commission (SEC) regulations—may require brokerage firms to preserve electronic communications relating to business for three years.

Box 8.2

CASE IN POINT: ACT IN HASTE, REPENT AT LEISURE

The very convenience of e-mail's send, carbon-copy (cc) and reply features are a double-edged sword, especially when it comes to responding to contentious issues and forwarding mails to groups. Unlike oral dialogue, where individuals can take umbrage under *'he said, she said'* e-mails leave a digital trace that is indelible.

I had first-hand experience of employee-relation issues arising out of the use of e-mails a few years ago when I was based out of Colorado. A close Indian friend of mine, who was an onsite co-ordinator for a vendor organization, had to bear the brunt of a cross-cultural misunderstanding. This gentle-man was in the habit of forwarding innate e-mails with jokes and innuendo among friends and colleagues in his 'personal' distribution list. On one occasion, he unintentionally crossed the line by sending out an attachment with a violent video stream.

This was just after the infamous '9/11' when Americans were already jittery over the state of security. One of the Americans in his distribution list, a project leader in our company, got per-turbed and instead of confronting my colleague, forwarded the mail to the HR department. Without any notice, my colleague was handed a pink-slip and escorted out by corporate security. It was unfortunate that he also happened to be on an H1 visa, which was also revoked and he had to migrate back to India.

His offshoring vendor company also had a huge fallout from this. Our IT manager lost confidence in the offshoring team and refused to renew any contracts; which also meant that the client's PMO would not consider their offshoring services or other projects in the organization.

Instant Messengers

Instant messaging (IM) technologies, which are practically free and provide instantaneous communication, are poised to provide connectivity in a networked world and are gaining in popularity among global teams. Combining the real-time advantages of a phone call with the convenience of an e-mail, IM is so compelling that it sometimes gets implemented through the back door, with distributed workgroups downloading public IM clients and using them before getting the nod from IT departments. Businesses are encouraging the use of IM tools to speed-up and ease communication. For instance, at IBM, some 220,000 employees worldwide are registered for instant messaging. Users can search in-house experts on a whole range of topics and requisition their expertise at any time.

IM, like e-mail technologies, uses the Internet as the underlying framework. However, unlike e-mail, IM systems are close-ended; MSN (from Microsoft), Yahoo and AOL use proprietary messaging techniques and contact lists, are generally close ended and do not talk to each other. A user of MSN Messenger software needs to install the package on his/her machine, signup for a login ID and use it to communicate with other MSN users. By doing so, they cannot use an MSN account to communicate with a friend who has an AOL or Yahoo account. Some of the key imperatives to be considered by managers wishing to advocate IM in their teams include:

- *Security:* Information Security remains a prime management consideration in implementing tools of open communication. Therefore, corporations generally try and discourage the use of 'open' systems like MSN or AOL. Instead, they prefer to encourage products like Lotus Sametime8 that can run within an organizational firewall. For instance, the US Navy uses the encrypted version of Sametime to help sailors communicate at sea.
- *Policies on the use of IM:* IM is a compelling business tool and its use can lead to an increase in productivity; however, its misuse may also inadvertently expose of sensitive business information. Many companies that have started documenting a detailed corporate electronic system use policies including sections on IM and e-mail.

IM is a synchronous communication media requiring both parties to be online; as teams may be spread across time-zones, managers need to encourage judicious use during commonly agreed times. As the global village gets more integrated, individuals are beginning to expect instant communication as a way of doing business.

Blogging and Wikis

Like e-mail and instant messaging, internet based technologies of Blogging and Wikis are catching the attention of the corporate world. Blog is a short form of Web logging, basically an online journal where the author (the blogger) keeps a running account of whatever s/he is thinking about. The blogger posts a paragraph or two on any topic every day and may even weave hyperlinks to other websites from the text. Business executives, especially senior managers in the corporate world, used to the 'old school' of thinking where they managed to control the flow of information in and out of the organisation, are trying to come to grips with blogging and instant messaging. Managers became interested in this technology after the publication of a case study in *Harvard Business Review*[9] that draws attention to the subtle role blogging can play in marketing and communication.

Managers are also beginning to realize the implication of blogging as a tool to foster team building and to help with information sharing. Team and project blogs are beginning to replace newsletters and e-mail-updates since information can be posted and shared across the globe almost instantly. Maintaining blogs also help managers publish their diaries, calendars and schedule effortlessly. Minutes of meetings and documentation of significant events can be recorded along with notes on status, changes in technologies and all other issues pertaining to a project. A properly managed blog can complement formal and informal communication channels and provide a 'water cooler' like environment for project teams. A case in point is the 'Blogs to manage projects' from Wilcox. (Ref: Box 8.3).

Box 8.3

CASE IN POINT: BLOGS TO MANAGE PROJECTS

Wilcox Development Solutions[10] is a small company based in Mansfield, Pennsylvania that provides Macintosh consulting for a wide variety of data related businesses activities. The company has deployed an innovative 'Project Blog' (Ref: Fig 8.3) system and allocates individual weblogs to help customers keep track of projects. Each project is given a code-name and a category so managers and customers do not have to look at information for all projects. The company's founder, Ryan Wilcox articulates the benefits of the weblog:

BOX 8.3

CASE IN POINT: CONTINUED...

"Using the project blog is certainly an interesting way to keep clients updated. While I use it more for Research and Development rants and learnings, I'd like to think that the project blog is useful for my clients as well.

It becomes more important to me now that I see e-mail eroding away—spam and other e-mail abusers being the biggest reason. While a customer has to go to my site to read the blog, at least I know that an entry can be seen, and not filtered away in a mountain of spam. Also, if there are 5 people on the project, it's a good way to keep everybody on the same page (without having to CC an e-mail 5 times, etc.)"

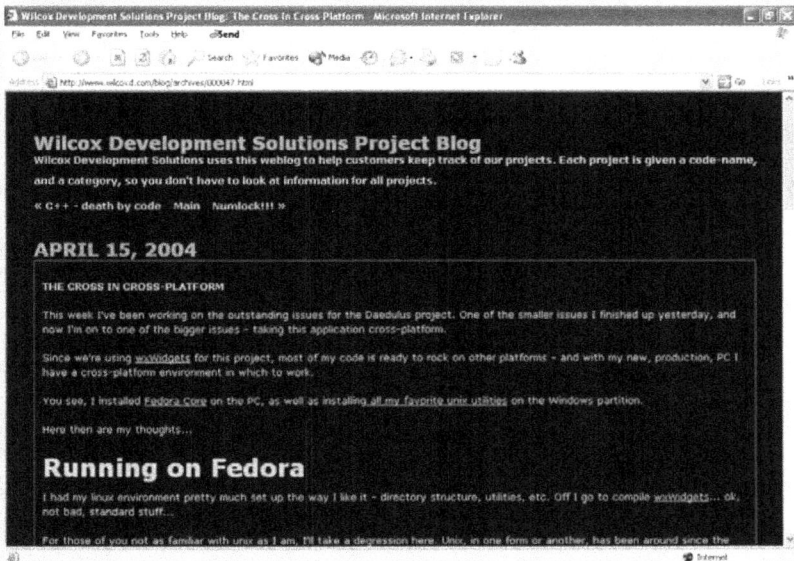

Fig. 8.3 Wilcox Development Solutions Project Blog

Video/Audio conferencing

Technical people are comfortable using collaborative tools in their workplace; for instance voice conferencing and the use of Voice over Internet Protocol (VoIP) phones has really taken off and facilitated cheap cross-border communication. Innovations in telecom technologies have not changed the fundamental nature of voice communication in years. Although telephonic technologies including voice conferencing has come a long way since its invention by Alexander Graham Bell, the basic usage is still a point-to-point voice communication, the protocols of which are very well defined. Audio conferencing supplemented by other offline communication models continues to be the backbone of business communication and a prime driver in offshoring. Most large service companies have installed their own dedicated links using VoIP or satellite links to mitigate the cost of voice messaging and conferencing.

Managers, sales people, consultants and specialists who generally prefer to conduct most of their business face-to-face with customers are starting to rethink their strategies and there is renewed interest in teleconferencing technologies. What is the next best thing to a face-to-face meeting? The obvious answer is video/teleconferencing. The technology to facilitate video-conferencing and tele-conferencing has been around for a while but widespread usage hasn't followed. Why hasn't video-conferencing become popular?

The reasons are not hard to find. People are generally more comfortable in face-to-face meetings. They like to observe body language and other non-verbal aspects of communication that are not very easy to capture on a screen. This is especially true if one is sitting in a group and the camera is focusing on only parts of the group. There is another human element to video-conferencing in groups: the camera will generally focus on the group leader or the senior most member of the team. Hence the purpose behind the conference—to see and observe—is wasted. It is especially difficult to grab the attention of all attendees all the time; added to this is the possibility of being disturbed, called or paged during video-conferences, something that the party at the other end will be unable to appreciate. Carmen Egido[11], in a paper reviewing the failure of Video-conferencing says, *"The experience to date, however, yields increasing evidence that video-conferencing is not the communication mode that lies between the telephone call and the face-to-face meeting, and that there are few examples of travel substitution directly attributable to video-conferencing or, for that matter, tele-conferencing in general."*

Though we have highlighted the pitfalls of video-conferencing and the reasons for slower adoption, managers are definitely looking at emerging practices and trends. There is also a cost advantage to look for an alternative to travel; a typical business trip undertaken by a team of two mid-level managers can set a company off by about eight to ten thousand dollars. Video streaming and webcasts (multicast, broadcast etc.) are increasingly gaining popularity in the education space and have applicability in discriminating project information, knowledge sharing and in other niche areas. Even with slower adoption, videoconferencing can enhance social relationships by putting a *'face to the name'* and allow better project co-ordination.

Other Technologies

The success of globalized project management hinges on the use of tools and technologies to manage projects, interactions, communication and the workflow. Tools and techniques that can streamline the interaction between offshore and onsite teams and mitigate the challenges of time and space are being adopted at a fast pace; not surprisingly, innovative adoption of tools and technologies in managing globalized projects is already gaining prominence among practitioners. Echoing these thoughts is a remark from John Tuman[12]: *"In the global arena of the twenty-first century, corporate strategy will be shaped by technology-empowered project management and a new generation of project managers that have the skills and the aplomb to apply modern tools and techniques to complex undertakings."* Tools including groupware, meetingware, Know-bots: (intelligent robots), search engine robots and digitization of documents are just some of the innovative uses of technologies in planning and managing projects. Project management software vendors have recognized this niche in globalized project and workflow management and have begun offering extensions to their applications for this. A case in point is the remote progress tracking and monitoring of milestones, tasks and activities facilitated by a collaborative feature of Microsoft's Project[13] (Fig. 8.4).

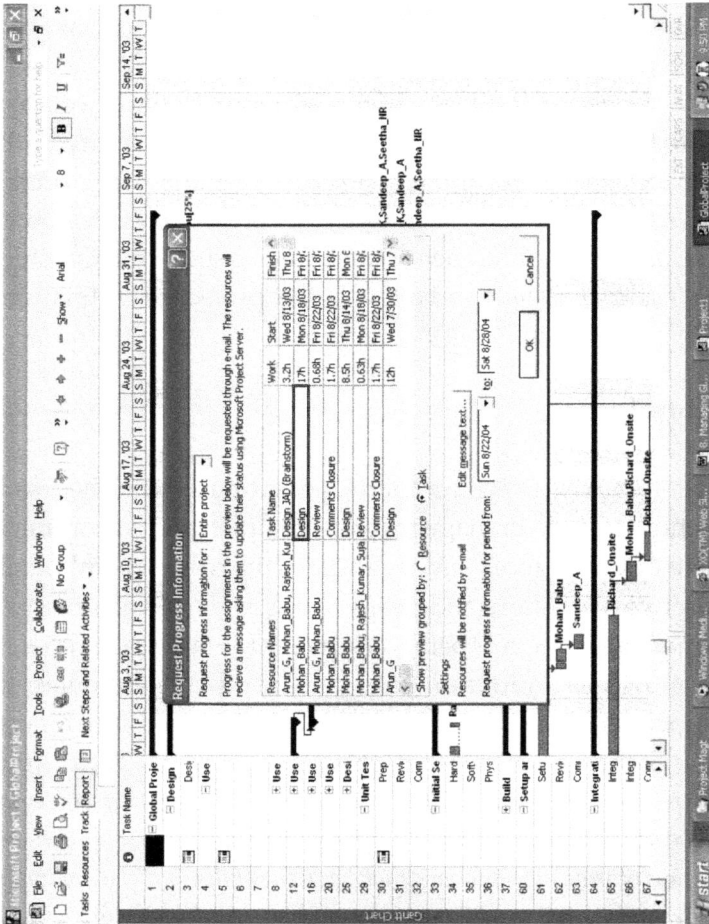

Fig. 8.4 Remote progress tracking using the collaborate feature of MS Project

Managing the workflow of application development and tracking is also gaining prominence along with globalization of project management tools. Software and service organizations are constantly attempting to link most of their systems and applications to provide their knowledge workers and managers with a unified view of the organization. Knowledge portals and Intranets are being adopted to aid faster and more accurate decision-making. In the earlier discussion on Project Management, we examined Infosys' tools and processes. Similarly, offshoring services company Wipro[14], has an integrated suite of tools to facilitate project management: *"The Project Data Bank (PDB), an integral part of KNET, contains detailed information about closed projects. PDB is a regularly updated repository of all the projects by Wipro. The integrated process automation tool (iPAT) is the project management tool used to manage and capture all project related material. The PDB gets direct input from the iPAT once a project is closed. All employees have access and can refer to the PDB for knowledge and experience gained from previous projects."*

Collaborative tools can help streamline the management of workflow, automating administrative tasks like gathering status reports, updating milestones etc. Although tools and technologies have a significant role to play in offshoring, success still depends on the management of teams with the right perspective, as David Pells[15] says *"Global corporations need project managers and project team members who have international experience and often global perspective. And more multinational organizations and projects increase the need for global project management standards, certifications and co-ordination within the global project management community."* Extending this is the viewpoint on offshoring communication (Ref: Box 8.4) that focuses on aspects of planning and strategic management of offshoring. Research in the field of knowledge servers, multi-use ontologies, knowledge bases and knowledge systems are already being field-tested by service organizations. A well defined project using the right tools can function as a manager's dashboard.

Box 8.4

VIEW POINT: OFFSHORING COMMUNICATION

When companies engage in offshore outsourcing, they don't think about *how* to communicate it until after the global switch has been turned on. And by that time, employees, customers, shareholders and media may have already formed opinions about the outsourcing program and, by extension, the company—opinions that are often times formed from incomplete or inaccurate information. The advice to executives and offshoring planners is simple:

First, make sure you have a fully-engaged executive sponsor who is willing and capable of serving as the voice for the company on its offshore strategy. This will usually be the CIO or CTO. He or she needs to be able to present the case for offshoring well to multiple audiences—most importantly, employees—and may need training. Don't be bashful about enlisting the support of people who can put them through communications boot camp on this specific issue. Much better to suffer the slings and arrows in private before opening one's self to the same scrutiny in public.

Second, any company that engages in offshoring that may cost people their jobs needs to understand it's a sensitive issue that should be communicated honestly, openly and with compassion. The key is to do so without appearing to apologize for offshoring in the process. Anyone who does that is going to wish they were in Bangalore instead of their programmers. Explain, communicate and provide rationale that speaks to the benefits of the program, but don't run from the issue.

Third, if ever there was a time for communicators to insert themselves into the operational planning phase of any program, now is the time and this is the issue. It's always surprising—not to mention disheartening—to see communicators on the outside

Box 8.4

VIEW POINT: CONTINUED...

looking in, only being called on after the offshore program has launched. Trying to understand all the nuances and complexity of an offshore program after it has been constructed is time consuming and ill-advised. Communicators need to be a part of the planning process long before the program is introduced to employees or outside audiences. We need to invite ourselves to the party.

Most importantly, once the strategic decision has been made about whether and how to offshore, all executives need to have the courage of their convictions. And this will require that we, as communicators, step up to the plate and start talking about offshoring the right way.

–Tom Phillips[17]

CONCLUSION

The focus of global managers is to synchronize the efforts of global teams and implement policies and the best practices of adopting the right tools and technologies. In our discussion of the Communication Layer of OMF, we focused primarily on the use of tools and techniques, highlighting the popular tools with the underlying assumption that the basics of communication management and the general body of knowledge on managing international teams, the cultural and language barriers and other factors will continue to be the foundation.

While managers, executives and global teams in the corporate world focus on aspects of communication, entrepreneurs are also positioning innovative training and solutions. Recruiting and retaining teams comfortable in a cross-cultural context is among the biggest challenges of offshoring management. Here is an example of a trainer providing 'accent reduction correct grammar usage' services, focused at offshoring service providers.

Box 8.5

CASE IN POINT: THERAPY/TRAINING IN ACCENT REDUCTION/CORRECT GRAMMAR USAGE

I am a speech/communication trainer with a masters degree from Columbia University. I have a private practice in Northern New Jersey. I provide therapy/training in accent reduction/ correct grammar usage for individuals who have relatively good command of English but are frequently misunderstood when talking with others due to a "thick accent". My services can be of particular help with business professionals who may feel they would be better able to advance within their career with better ability to speak and communicate effectively in English. Anyone living in the northern NJ area is welcome to contact me for further information about my services.

Judy

(Source: GaramChai.com[16])

NOTES

1. Dr. Shirley Says...You Cannot Not Communicate [Success Images, http://www.successimages.com/articles/sw01.htm]

2. *A Guide to the Project Management Body of Knowledge* [PMBOK® Guide from the Project Management Institute (PMI®)].

3. *Beyond Culture.* [Hall, E.T. (1989), Anchor Books Editions.]

4. *Cultures and Organizations: Software of the Mind.* [Hofstede, G. (1997). London: McGraw-Hill.]

5. Managing cross-cultural issues in global software outsourcing [S. Krishna, Sundeep Sahay, Geoff Walsham, *Communications of the ACM*, Volume 47, Number 4 (2004)]

6. Online resources listing cultural differences:
   http://www.immihelp.com/newcomer/language.html
   http://www.rso.cmich.edu/iso/american.htm

7. Are e-mails Worth the Paper They Are Written On? [By Gregory A. Nylen, Esq.; From *New Matter*, Volume 27, Number 1, Spring 2002 http://www.gtlaw.com/pub/articles/2002/nyleng02a.asp]

8. IBM's Lotus Sametime: Software vendors like IBM make encrypted versions of IM tools for use by the academia and military.

9. A Blogger in Their Midst. [September 2003 issue of *Harvard Business Review*]

10. Wilcox Development Solutions [http://www.wilcoxd.com/]

11. *Video-conferencing as a Technology to Support Group Work: A Review of its Failure* [Carmen Egido, Bell Communications Research, Inc.]

12. Shaping Corporate Strategy with Internet-Based Project Management [John Tuman Jr., *The Future of Project Management*, PMI]

13. Microsoft Project: Microsoft.com [Project]

14. Wipro Case Study; Knowledge Management Portal Saves Time, Money, and Improves Productivity at Wipro [Published on Microsoft.com http://www.microsoft.com/india/casestudies/wipro.aspx]

15. Global Tides of Change: Significant recent events and trends affecting globalization of the Project Management Profession [David Pells, *The Future of Project Management*, PMI]

16. Bulletin Board GaramChai.com [http://www.GaramChai.com]

17. Tom Phillips is managing partner of DDB Public Relations and Director of DDB PR's Outsourcing Practice Group headquartered in Seattle. Phillips counsels companies on the communications and image aspects of offshore outsourcing. DDB is one of the world's largest marketing communications firms with 200 offices in 99 countries.

CHAPTER 9

Managing Globalized Workforce

- 💻 Cultural Aspects of Offshoring
- 💻 Managing Technical Aspects
- 💻 Human Aspects
- 💻 Conclusion

Software development is an intellectual activity that cannot be accomplished without groups of skilled and talented people synchronizing their efforts towards a common goal. There are various facets of globalization and software development that encompass the Offshoring Management Framework that we examined in the four *Layers*. The layers and interaction between and across them focused on the process dynamics and interfacing of onsite and offshore teams. Offshoring also necessitates interaction, communication and networking with partners, vendors, suppliers, outsourcers and others from around the world. Managers and members of teams need to learn to work with professionals and peers from across geographic and cultural boundaries. The key challenge is to manage the dynamics of global workforce, also called *geographically distributed teams* or *virtual teams*. Some of the key aspects of managing a globalized workforce include:

- **Articulating clear goals:** The onsite and offshoring teams need to be clear about their goals and objectives. By defining the goals clearly, the Management can ensure that there is a common understanding among team members and there is a frame of reference to work towards the deliverables and targets.

- **Define modes of interaction:** Stakeholders across the offshoring spectrum may have varying goals and needs that will have to be addressed and planned for. Interaction between the layers of OMF, across onsite and offshore teams, and with stakeholders outside the performing organizations will have to be planned and articulated

- **Focus on communication:** Defining the modes of interaction may include communication management planning, essentially aiming to bridge the onsite-offshore gap. This may also encompass the usage of tools and techniques, bridging cultural and language barriers.

- **Cultural differences:** Teams and managers may have to recognize the existence of cultural differences that may not be mitigated. Such recognition may dawn from an awareness and appreciation of the diversity of backgrounds, collective experiences and insights that individuals from other cultures and backgrounds may bring to the project.

- **Trust between teams:** Working with teams across geographic and cultural boundaries hinges on basic trust between onsite and offshore teams. Such bonding and build of trust should accrue speedily to be beneficial to projects of shorter duration.

Managers are responsible for facilitating the creation of globalized environments and ensuring that the team they are building appreciates the nuances of communicating with people across boundaries. While managing the different aspects of a software project, the aim is to bring together a group of people in different geographies from different cultures, with varying goals, skill-sets and experiences and ensuring that their work is orchestrated in a cohesive manner. In the rest of this discussion, we will focus on three key aspects pertaining to the management of global teams (Fig 9.1) including the human, technical and cultural aspects. To set the context for the discussion on managing a global workforce, a discussion of a 'typical day in the life of' follows. Although the references for these two cases have not been provided, they are typical of offshoring management jobs, especially in the service delivery management arena.

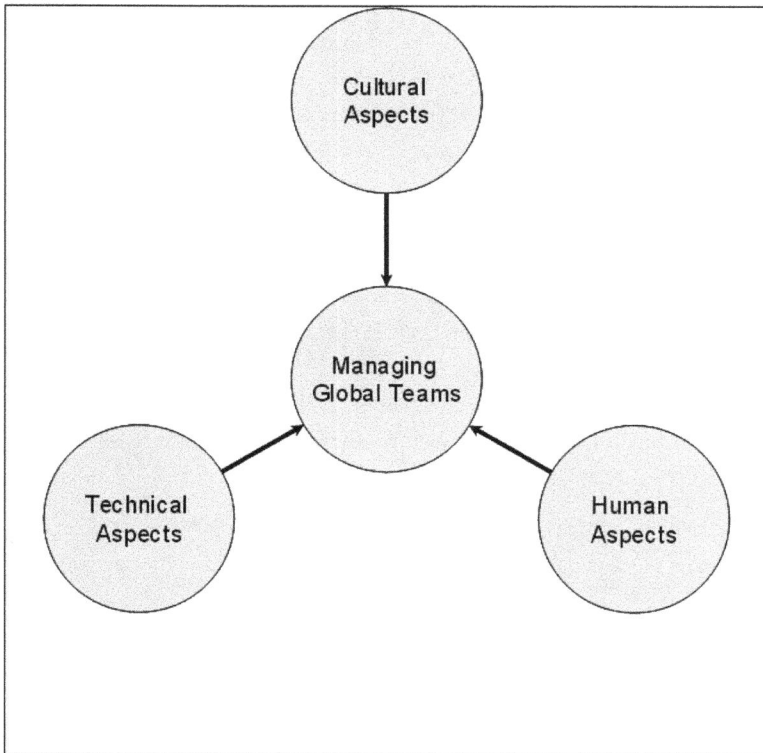

Fig. 9.1 Managing Global Teams

Box 9.1

CASE IN POINT: A DAY IN THE LIFE OF...

A Day in the Life of an Offshoring Project Manager

I am Harjeet Singh, Harry for short. I manage two projects for a Canadian bank out of the Hyderabad office of my company, a large offshoring software integrator. My typical day begins early, with a call from the onsite folks briefing me on the intricacies of the testing at their end. Development on one of the projects has just finished and our systems were recently delivered onsite for a systems integration test and a few issues are surfacing during the porting to the new infrastructure. My offshore team is diligently supporting the issues but some are getting escalated to me and the management, requiring more attention.

I reach the office at about 9 AM and after a coffee at our canteen, settle into meetings and calls. Today we have a prospective client visiting us and I need to give a presentation on our Project Management processes. The presentation goes off well since most of their questions and queries are those we have faced from other customers, and includes the 'how to' on global delivery, a demonstration of our processes and tools at work. Another major agenda for the day is a planning meeting where the Delivery Manager wants inputs on the resource forecast for the next six months to a year that will have to feed into our company's annual planning. Before leaving work at 7:30 PM I catch up with a review of code and charges for tonight's code-drop.

Back home, I attend to a few conference calls before calling it a day.

A Day in the life of an Onsite Co-ordinator

I am Nina, Originally from the Indian "Mid West" where I did my degree in engineering from a top-rank university after which I joined a "Billion Dollar" Indian IT giant. This was a

Box 9.1

CASE IN POINT: CONTINUED...

coveted job and I was picked right out of campus and sent to a three month long boot camp where I learnt the nuances of application development and delivery. After spending a few years working on offshore aspects of project execution, including offshore project management, I moved to the American-Mid West where I find myself co-ordinating projects my company is doing for this client, a large Railway carrier.

My day typically begins early at about 5:30 AM. After a hot cup of tea, I login to check my mail and chat with the offshore team over instant messengers; calls with the offshore team and managers over the day's deliverables may supplement the online chat. Calling our Bangalore office is simple; I just dial a 1-800 number (Toll free) and am connected to our office via a dedicated satellite link. My conference-calls and mail-checking is frequently interrupted by calls on my cellphone from the account manager and other folks.

I generally wrap up the morning session by about 8:00 AM and head to the clients office to begin my 'day job' of being an onsite-co-ordinator. After a quick meeting with the onsite technical teams from my company I get busy with my daily routine that includes verifying code-drop from offshore and ensuring all the deliverables the offshore team sent the previous evening are compiled and tested. Between client meetings and other brainstorming sessions, I manage to break for lunch where I catch up with onsite folks from our company. The meetings and analysis sessions continue till late in the evening, sometimes involving conferencing with our offshore team members during their early mornings.

I usually head back to my apartment at about 8:30 PM and after dinner, settle down to catch up with the offshore teams again, this time to do a brain-dump of my client interactions and set the priorities for their day. Other aspects I get involved in include—account management, billing, presales, working on responding to proposals with the offshore team and also give the project management team inputs on work allocation, scheduling etc.

Typical tools of my trade include my Blackberry, cell-phone, laptop and PDA. Once I am online, I depend heavily on the in-house tools of my company including the workflow, project management, proposal management, CRM, e-mail and other systems. The internal knowledge management system helps me gather templates, documentation guidelines and other artifacts. Although my job is really hectic and there are times I feel I am on a tread mill, torn between demands and challenges of two worlds—literally—I would not trade the experiences and learnings of this job for any other.

CULTURAL ASPECTS OF OFFSHORING

Aspects of culture and communication have a significant role to play in offshoring application development although the impact of such aspects can be minimized by documenting the application requests in a standardized way. For instance, development of core algorithms to be used in a search engine, or development of software to be embedded in devices can be specified in a relatively culture-neutral way. Application development, on the other hand, depends heavily on customizing inputs on usability, business domains and verticals that may vary based on geographies; therefore it may need a stronger focus on cultural and communication aspects. Development of business applications, for instance, increase the need for interactions with users for definition of the specifications and requirements.

Typical application development endeavors require people with varying skills—technical writing, usability specialists, testing, marketing, domain management and project management. Projects, programs and engagements between clients and vendors may also require interactions between executives, planners and other business leaders. Stakeholders at an offshoring initiative may also need to interact with the onsite and offshore teams. Interactions between and outside project teams will require attention to cross-cultural aspects including focus on communication, language, conceptualization of solution and ideation. Offshoring also brings about an increased need for mobility of individuals who are comfortable in global settings. The mobility of Indian professionals, which is among the key success factors of Indian offshoring phenomena, has not gone unnoticed by management gurus and thinkers (Ref: Box 9.2).

Box 9.2

FROM TUMKUR TO TORONTO

During a seminar on offshoring and globalization that I attended, Prof. C.K Prahalad gave a keynote address where he talked about some of the key drivers in the industry. Among the points he highlighted were the usual suspects of the emergence of better educated, more mobile and culturally aware workforce. An example he quoted stood out.

While explaining the tenacity of Indian professionals, he alluded to the fact that the real edge of people from India and other developing economies moving to the west to participate in global projects was their cultural adaptability forced by the economic disparity between their home countries and the client countries. Prof. Prahalad gave an example of *'a kid fresh from engineering college in a small town, say Tumkur, in South India'* more than willing to relocate to any corner of the globe

> ### Box 9.2
> ## FROM TUMKUR TO TORONTO
>
> with minimal lead time. The *'kid,'* said Prahalad, needed little cultural re-orientation or insights, and was motivated enough to travel with just his passport stamped with a visa, a few technical manuals, the address of the motel and client and some traveler's checks.
>
> On landing in a faraway place, say Toronto, the techie from Tumkur not only finds a niche at his new workplace but also uncannily finds the nearest eatery serving Dosa (a south Indian entrée). Try getting a multinational to do the same with a 'kid' from Ohio and place him in Bangalore, Prahlad challenged the audience. The example quoted by Prof. Prahalad is being played out hundreds (if not thousands) of times every day with Indian professionals criss-crossing the globe. Several specialized Web-Portals and information directories catering to the Diaspora like GaramChai.com[1] have become popular in this space, as have blogs and online discussion boards. Interestingly it is not an Indian phenomena alone; with the opening up of the European Union and the addition of newer East-European countries, mobility of professionals is a key success factor behind offshoring.
>
> –Story idea extracted from the Author's Column[2]

Jobs and projects are increasingly being executed across borders, which also means that individuals need to learn new skills of multi-culturalism. People across the globe embarking on new careers are realizing the significance of cross-cultural workforce dynamics and are making conscious efforts to gain a heads up. Though management gurus and business leaders emphasize the need for greater cultural awareness, it is the individuals who are now taking the lead by preparing to be a part of the global marketplace. Some of the major issues on aspects of globalization and cross-cultural work include:

- **Managing cross-cultural teams:**   Managing global teams is gaining as much prominence as is working with cross-cultural teams. Work teams and groups are getting to be really international, especially for people in the software industry. As a colleague, manager or supervisor, one will have to identify with teammates, some of whom are going to be from a different cultural and ethnic background, and ensure that they are on board with regards to the objectives of the team, organization and group.
- **Language:**   English has become the language of globalization; for instance, most of the IT projects and systems moving to India are from English speaking western nations. However, there are still large pockets of the world where English is not the Lingua Franca. For instance, Benjamin Limbach[3], A researcher in Europe says, *"there is a hype in Europe at the moment to outsource projects to e.g. Poland, Hungary, the Czech Republic, etc. due to short distances, extreme salary differentials and highly qualified manpower; more so in the context of offshoring initiatives, focused on the 10 countries that joined the European Union as of May 1, 2004. The majority of these countries are in the east of Europe hence they call it 'nearshoring' here in Germany."* The language barrier can somewhat be mitigated by employing translators and bilingual team members but lack of expertise in a language can be a big deterrent in offshoring initiatives. This is probably the reason why more projects from European nations are being *nearshored* to East-European countries, not to India or elsewhere.
- **Abundance of 'best practices':**   Information on cross-cultural sensitivity, cultural nuances etc are abundant and include books on culture and international travel and etiquette, internet and training courses. Large organizations

employ consultants and specialists who coach staff on aspects pertaining to international interaction. Even universities are beginning to offer courses on international management, cross-cultural sensitivity and globalization. Best-practices supplemented by hands-on exposure to other cultures by selective job-transfers and travel can help teams synchronize and work together.

- **When *not* to go by the book:**   Most textbook approaches to managing in a global environment may be highly dependent on specific contexts; and some of the 'cultural nuances' highlighted and underlined in management and travel books are a bit over-stated. For instance, the typical English stiff-upper-lip and American boorishness, which is a part of several books on globalization, is a bit overrated. I have dealt with a lot more boorish Indians and stuffy Americans than the textbooks would acknowledge. The fact of the matter is that foreigners are given extra leeway, especially when it comes to subtleties in cross-cultural interactions; as long as they don't commit any obvious faux pas. General cultural traits may surface at different times, and those are to be recognized and dealt with.

- **Subtleties of cross-cultural work and communication:** Though there is an abundance of information and data on 'best practices' pertaining to cross-cultural interaction and working in globalized environments, there are nuances to such interactions that only come with experience and insight. For instance, my experience is that the importance of making 'small talk' is underestimated. In some cultures, especially in the west, it is by small-talk and during water-cooler conversations that individuals gauge others before and during meetings in work and cultural settings. Talking about one's football or soccer team, for instance, is a way of bonding with co-workers, indirectly asserting that there is life outside work. (Ref: On a personal note, below) The Authors in a whitepaper[4] add: *"For example, Indian software companies have* *found they need to approach communication with US and Japanese clients in very different ways. US client companies normally work with extensive written agreements and explicit documentation, reinforced with frequent and informal telephone and email contact. In contrast, Japanese clients tend to prefer verbal communication, more tacit and continuously negotiated agreements, and less frequent but more formal use of electronic media."*

- **Team Dynamics:** Experts who study organizational behavior agree that teams and groups build their own working dynamics. As a colleague, manager or supervisor, there may be steps that one might have to take to identify the important aspects of each culture in your department or within the team. I have known western managers who attempt to draw out their Indian colleagues by getting them to talk about the sub-cultures of the Indian subcontinent, languages, ethnicities and vegetarianism. They do it consciously without being abrasive so that their foreign colleagues get a sense of belonging and bond with the team. Similarly, I have known Indian colleagues take their British or German co-workers out to local pubs during their stay in Bangalore. Though such 'clubbing' may not be in any *job description*, it definitely improves camaraderie. Another example of team dynamics in a globalized team was given by the authors[4] of a paper who say that *"British managers in an outsourcing relationship with a particular Indian software supplier found that Indian programmers, in deference to authority, would not voice criticism in face-to-face meetings but would sometimes send their opinions in email messages after the meetings had disbanded."*

- **Leveraging the software culture:** Software professionals, like most other engineers have a strong allegiance to the field of software engineering. Erran Carmel[5] argues that, *"Software professionals worldwide belong to the computer subculture… Software guru Larry Constantine argues that the computer subculture is stronger than national culture and that the program-*

mer in Moscow is more similar to his American programming peer than to other Russians.... Engineers, like software professionals, place high value of on work and on achievement and relatively low value on social relationships. The stereotype of the antisocial programmer has a kernel of truth." The author goes on to successfully argue that though working in multicultural environments requires continuous learning about cultures, *"fortunately, our common (professional) software culture is a unifying force."* Recognizing the existence of such a software culture is a key step for managers to build a cohesive team culture that can tide over smaller challenges posed by regional or geographic subcultures.

IT managers are coming to realize the significance of cross-cultural communication. Organizations are enabling teams and managers to become more *global* and take on challenges of working with teams from across geographies and cultures. Though there are several theories and practices on managing offshore teams, the actual process has a strong human management angle. To successfully manage offshore and onsite teams managers need to innovatively foster team-building. For instance, there may be a need to let team members know that they mean more to projects than an e-mail address or a teleconference voice. Personal events may be hard to recognize and 'celebrate' with remote teams but an e-mail greeting or a mention during a weekly status call can make a big difference.

The offshore application development model hinges on reducing costs by managing people in different corners of the globe. This also implies that project plans will attempt to minimize cross-country travel during the course of the development life cycle. Due to budgetary or other logistical constraints, a manager may never be able to get the entire team together in one room. For instance, I have personally managed projects for clients in US and Canada where I did not travel onsite to either visit the client or meet with the onsite team. In such situations, the project manager needs to be extremely empathic towards members of team who are located distantly.

ON A PERSONAL NOTE

I was based in the beautiful Wewlyn Garden City in Hertfordshire, England during the mid nineties. This was one of my early work-assignments as a consultant after graduating from college and I was as green as they came. My employer, in the true spirit of 'body shopping'—as was the norm then—sponsored my visa and packed me off to the UK with some foreign currency and reference to another senior from the company I was to relieve.

The Indian senior I was to relieve focused on the technicalities of the assignment for the week that we had a handover and then left. It was now up to me to understand the cross-cultural nuances and begin working with my new boss and colleagues, who were all British. It helped a bit that they had experience working with Indians, especially with consultants from my company who regularly visited the UK.

Chronicling my experiences during the first few months living and working in England would require a book by itself but one aspect of my stay stood out. The team I worked with had a 'culture' of going out for lunch to the local pub every Friday where either the manager or one of the team-mates would sponsor a round of beer. I had been told that many of my Indian colleagues who had worked there previously had refrained from joining the team during their outings. I had no such inhibitions and began going out with the manager and my colleagues. The first thing I noticed was that more than going on a drinking binge, the act of walking to the pub with colleagues was a bonding ritual. All the topics under the sun, including aspects of our project and work would be routinely discussed in an informal manner.

While going out to a pub with colleagues is not the only way to bond, it was certainly a custom followed in that group, and participating in that group *cultural* activity helped me bond better.   *–Author*

MANAGING TECHNICAL ASPECTS

Business executives and leaders are increasingly becoming aware of the need to include IT in the strategic mix and planning. They are actively getting involved in IT initiatives instead of passively observing their technology managers recommend solutions. On the other side of the spectrum, IT managers and executives are also beginning to come to grips with this shift and the fact that their existence depends on the success of the overall business strategy of the company and business units they support. Offshoring is bridging the business–technology divide since, by offshoring aspects of the development cycle, technology managers delegate the technology management aspects to offshore partners while focusing on business issues of technology management.

Traditionally, IT Managers didn't get invited to strategic decision making meetings, a forte of business leaders and executives. However, with offshoring on the rise, the percentage of strategic influence that technology managers wield is increasing in focus. As can be observed in Fig. 9.2, technologists and IT managers traditionally focus predominantly on the technical areas that include architecting solutions, building infrastructures, application development, and less on the business domains that the solutions address. Traditional inputs

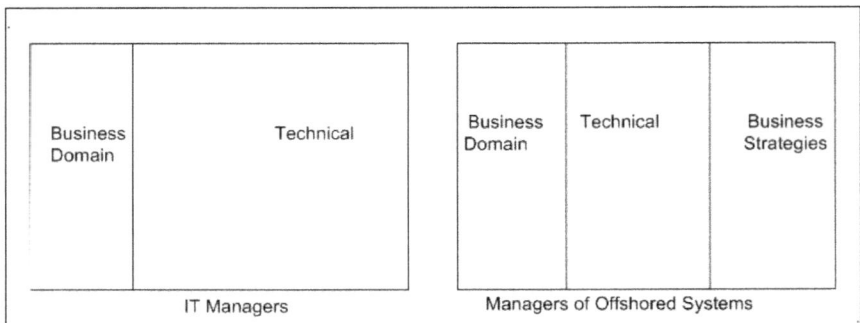

Business Domain	Technical		Business Domain	Technical	Business Strategies
IT Managers			Managers of Offshored Systems		

Fig. 9.2 Areas of focus for Technologists and Project Managers

from Project Managers and IT managers have included planning and operations of Application Development and Maintenance. By offshoring technology management and development, Managers are expected to bridge the gap between application development and the strategic goals of business. This focus now involves a shift upstream towards bridging the strategic needs of businesses, by suggesting innovative uses of technologies. Technology leaders are being expected to increase focus on strategic issues, while they continue to bridge the gap between business leaders and technologists.

The focus of offshoring projects is to ensure that the technical aspects of application development are abstracted and seamlessly handed over between onsite and offshore teams. In this section we will examine a few niche areas as they pertain to management of software professionals:

- **Team Culture:** Groups of individuals who come together to work, begin to share ideas, aspirations and build sub-cultures around successes and failures. Sharing common anecdotes and 'war stories,' joking, jibing and humor or in some cases, common disdain for certain processes or methodologies can all bring people closer. Managers need to recognize the existence of such cultures within teams and either participate in them or tacitly encourage formation of such group cultures. Team building is harder in an offshoring context where traditional techniques like team outings may be harder to implement. Tools like 'virtual' water-coolers implemented with blogs, chat-boards, discussion boards etc. may help foster team cultures.
- **Motivating individuals and teams:** Motivating includes aspects ranging from employing the right people for the right jobs, paying the best salary and packages and building the right work environment, down to identifying individual motivators. Motivation need not be about money. Managers of offshore and onsite teams need to acquire an acute sense of

managing across borders. For instance, simple things like sending a 'kudos' e-mail after a crucial software rollout can also go a long way in motivating individuals and making them feel a part of the team. Other techniques like recognizing variations in the tone of voice during a phone chat may supply clues that something may not be right at the other end.

- **Managing 'B Players':** Offshoring relies heavily on B-Players, all-rounders who cannot just code well but can also appreciate the nuances of communicating across geographies. Managers who deal with programmers and software professionals might recall working with one or more *'star coders,'* and are in awe of such super-programmers who can switch between languages, platforms and technologies with ease. It is also a fact that 'star coders' are really rare and hard to find, so most managers make do with *B-Players*[6]. IT managers realize the importance of their B Players, non-star programmers people who have worked on the systems extensively and are able to provide dependable, consistent output. This concept of managing B-players is especially useful for managing global teams as the attributes and traits of people and team members who come form different cultural backgrounds may not be very apparent during regular interactions. Managers will have to make a conscious effort to ensure that the B-Players who work hard to bridge the cultural divide in addition to taking on technical responsibilities, essentially being *'all rounders'* are recognized and motivated as much as the star performers.

- **Managing People:** Managers need to learn to diplomatically manage programmers and Subject Matter Experts (SMEs) who may know more than them; this is especially true of managing globalized workforces. The level of obsolescence in the field of technology is very high and managers who may have been programmers in the past find it hard to comprehend the 'nitty-gritty'. Another technique that managers find helpful is by scheduling periodic reviews of deliverables by external SMEs from other groups within the organization. This may help create the necessary checks and balances without undermining the programmer or architect's credentials. The same technique can be extended to onsite–offshore reviews and technical brainstorming sessions.

- **Managing across departmental boundaries:** Large service delivery organizations provide a uniform front to clients who generally deal with the onsite manager for most day to day interactions. We observed how onsite and offshore co-ordinators are the crucial link in bridging the offshore–onsite divide at execution time. (Ref: Box 9.1—A day in the life of…). The manager may have to interface across departments or groups to address various administrative and technical issues faced by clients. Ensuring that the finance department raises a proper invoice and bills the client for all successful milestones is an example. Other administrative issues could include managing the logistics of travel and ensuring that the team has all the software and hardware required for their jobs.

- **Knowledge sharing culture:** Offshoring managers need work hard to build knowledge sharing cultures in teams. In order to do so, they need to recognize the existence of two distinct kinds of software professionals: those who willingly share knowledge in all circumstances and those who believe in the dictum 'knowledge is power.' Individuals from the second camp come with a misconception that they will be diluting their presence in the team by sharing their knowledge of tools, techniques and best practices. Addressing such concerns and ensuring that such individuals feel a part of the team is the fist step to bringing teams towards a cohesive unit. One simple technique to foster a knowledge sharing culture among techies is to leverage the *software culture*. Software professionals generally exhibit a greater sense of camaraderie among others with a similar passion, and such affinity can span geographic and cultural boundaries. For instance, solving a 'bug' in a .Net code would thrill most good programmers working on Microsoft platforms whether they are Indian, Russian or American; managers can work on simple techniques to leverage the existence of such *software cultures* and foster a culture of knowledge sharing.

Technical people, including software professionals build their distinct work cultures, and managers need to understand the subtle nuances of managing technical people. In many cases it is easy since the managers themselves might have come up from the ranks. The 'software culture' among technologists permeates even ethnic, regional or national cultures. This can be leveraged by managers who can emphasize the software culture as a binding force among the team. IT Project managers constantly innovate on this aspect by creating myths about problem solvers and by turning failures into team-lesson sessions. Extending some of the key points mentioned above, communication and offshoring experts have also articulated best practices. As a case in point, Deena Levine has articulated some tips for offshore–onsite team building (Ref: Box 9.4).

Box 9.4

COMMON SENSE FOR OFFSHORE-ONSITE TEAM RELATIONSHIP...

...BUT NOT ALWAYS PRACTICED!

- Focus on how team or project success will be a win for everyone.

- Consider the human element—no matter how sophisticated your processes are, remember that human beings need to connect and communicate in order to create success in the project.

- Assume positive intent on both sides of a global team.

BOX 9.4

COMMON SENSE CONTINUED...

- Apply more rigor and discipline to all team processes than you might need to if the team were co-located and homogeneous.

- Set realistic timelines; solicit team members' input to determine what is realistic.

- Promote the collaborative model at all levels of the project; this style is more likely to contribute to a successful project.

- Create a collaborative team environment, one that carries over after team members return to their development centers.

- Check in with other team members about your own language use. "Am I speaking too fast?" "Would you like me to repeat it?" Let people know that you won't be offended if you are asked to repeat or slow down."

- Over-communicate and over-clarify—the communication process across cultures has more layers of complexity and requires extra attention, patience and time.

- Focus on identifying all stakeholders and keep people in the loop as much as possible (i.e., promoting inclusive team behavior as a prerequisite for global teams).

- Foster an environment in which there is open communication between managers and direct reports. Thoughts that are not expressed today may result in a loss for your organization.

- Don't jump to conclusions about teleconference behavior based on your initial interpretation; you may be misunderstanding cultural cues

- Focus on developing trust; build—in face-to-face opportunities, especially at the beginning phases of the project

–Deena Levine[7]

HUMAN ASPECTS

Managing a globalized workforce, essentially teams from across cultures and from varying backgrounds requires a deep insight into what really motivates people. Although aspects of culture and ethnicity can determine aspirations, at the very basic level human goals and wants continue to be similar. There are several *fundamental* Human Resources management theories that are perennially referred to. For instance, few discussions of people management are complete without a reference to Maslow's Hierarchy of Human Needs. Also of interest in such discussions is Frederick Herzberg's[8] study of what motivates people to work. Abraham Maslow[9] was a humanistic psychologist who believed that humans strive for higher levels of capabilities and seek frontiers of creativity and highest reaches of consciousness and wisdom. Managers can instinctively relate to the theory presented, especially the depiction of human wants in a pyramid format. (Fig 9.3)

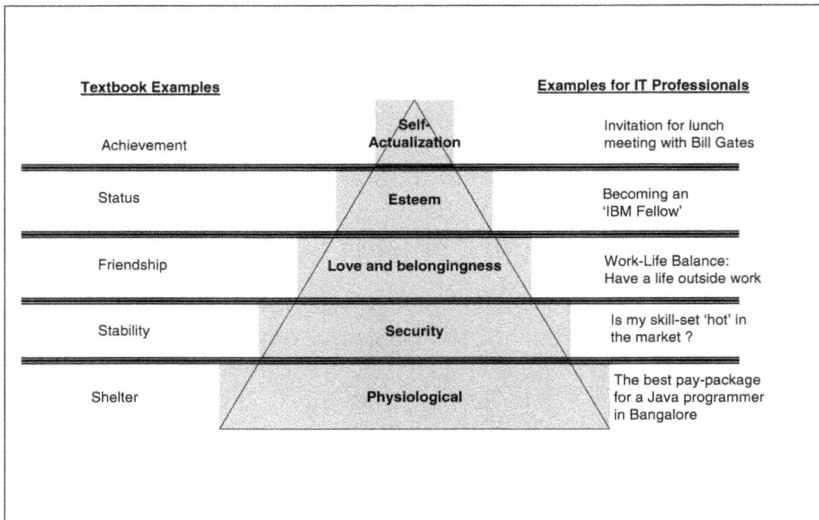

Textbook Examples		Examples for IT Professionals
Achievement	Self-Actualization	Invitation for lunch meeting with Bill Gates
Status	Esteem	Becoming an 'IBM Fellow'
Friendship	Love and belongingness	Work-Life Balance: Have a life outside work
Stability	Security	Is my skill-set 'hot' in the market ?
Shelter	Physiological	The best pay-package for a Java programmer in Bangalore

Fig. 9.3   Maslow's Hierarchy for IT Professionals

An awareness of the basic theories and factors affecting job satisfaction is essential for the management of individuals. Such awareness should translate into actionable practices that managers can apply in their projects and teams. Although there are cross-cultural and global implications of management of individuals, the basic theories of human motivation and management continue to be for the basis of managing global teams. Case in point: Architects of the open source movement realize that the core motivators for techies across the world who participate in ventures like SourceForge is an *Egoboo*, basically recognition among the peer group. (Ref: Box 9.5).

Box 9.5

CASE IN POINT: EGOBOO AS A MOTIVATOR FOR THE GLOBAL TECHIE

The open source movement is a phenomenon that is catching the attention of technology and business leaders alike. Business leaders are particularly intrigued by the fact that much of the open-source software development is done by highly talented professionals in their *spare time* and is given away *free* (under one of the open source licensing mechanisms) to the community at large. The software systems include the popular operating system Linux, web server Apache and database MySQL that are giving established organizations like Microsoft and Oracle a run for their money. The sophistication and complexity of such open source software developed by teams of motivated programmers from across the working in 'virtual' teams is perhaps something every project manager would dream of emulating.

The most fascinating aspect of the open source movement is that individuals are not motivated by money or payments; what they are really after is a 'virtual currency' called Egoboo, an abbreviation of "ego boost." Techies participating in the open-source movement are more motivated by ego-boosting

Box 9.5

CASE IN POINT: CONTINUED…

and enhancement of their reputations among fellow techies than by the lure of money. It is hard to define where the term Egoboo originated but a book by Gerald Weinberg titled *The Psychology of Computer Programming* has a discussion on "egoless programming" where the author observes that remarkable improvement in productivity can be seen in IT organizations where developers are not territorial about their code. This fact is also described succinctly by Eric Raymond[10] when he defines egoboo:

"The 'utility function' Linux hackers are maximizing is not classically economic, but is the intangible of their own ego satisfaction and reputation among other hackers. Voluntary cultures that work this way are not actually uncommon; one other in which I have long participated is science fiction fandom, which unlike hackerdom has long explicitly recognized "egoboo" (ego-boosting, or the enhancement of one's reputation among other fans) as the basic drive behind volunteer activity."

The eco-culture of open source is being studied in greater depth by management gurus and best-practices are yet to emerge. This also spells an opportunity for project managers and executives to think out of the box and experiment with innovative ways to motivate their technical teams.

In addition to motivational theories, managers of globalized teams need to focus on aspects of hiring and retention of workers around the world. This may include an awareness of HR management, legal and employee relations practices. Managers also need to be acutely aware of emerging trends in the way Computing Professionals view their lives and careers. The author[11] had articulated the transition of professionals from Organization Men to Free

Agents and argues, '*As professionals in a workforce with evolving expectations of the employer–employee relationship, most of us will need to acquire and apply entrepreneurial and business management skills to manage our careers. Our career trajectories will thus depend on constant marketing and networking rather than climbing the ladder of a predefined career track.*' A parallel trend shaping up is the emergence of Gold Collar workers in the global marketplace. (Ref: Box 9.6).

Box 9.6

MANAGING TECHNOLOGISTS: GOLD COLLAR WORKERS

Careers and professions as we know them are undergoing a remarkable transformation. Even as early as a decade ago, people joining large corporations were tacitly given to understand that the job was theirs as long as they were competent to work, i.e. till retirement. Most professionals joining large corporations would hope to gradually move up the ladder and eventually retire with a gold watch. This is not true anymore. With economies around the world becoming more integrated and market oriented, swings in economic cycles are prompting even traditionally stable companies to resort to layoffs, making jobs and career paths really unpredictable. The trend, of looking at a career as a hierarchical ladder is giving way to a new model—a career as a series of gigs. Some professionals have long been privy to this trend. Film stars move from one film to the next and athletes move from one game to the other, building experiences and marketing themselves. Similarly, professionals in most fields are slowly discovering that their career is also turning out to be a series of gigs. White collared professionals are realizing the need to manage their own careers without depending on one single corporation.

When asked if a young man in a gray flannel suit represented the lifelong corporate type, what today's image of a

Box 9.6

MANAGING TECHNOLOGISTS: CONTINUED...

corporate professional would be, Peter F. Drucker, godfather of modern management, replied, "*Taking individual responsibility and not depending on any particular company. Equally important is managing your own career. The stepladder is gone, and there's not even the implied structure of an industry's rope ladder. It's more like vines, and you bring your own machete. You don't know what you'll be doing next, or whether you'll work in a private office or one big amphitheater or even out of your own home. You have to take responsibility for knowing yourself, so you can find the right jobs as you develop and as your family becomes a factor in your values and choices.*". He went on to add "*.... whether you were in India or France, if you were an assistant director of market research, everybody used to know what you were doing. That's not true anymore....*"

In companies around the globe, layoffs, downsizing and rightsizing are becoming the norm rather than the exception. This does not mean that corporations will stop hiring talented and experienced people; on the contrary, there is going to be a renewed focus on attracting the right talent to get the job done. However, it means that most projects and jobs are going to become more finite, like gigs, at the end of which the employee will have to look for the next project, assignment or job. Professionals are realizing that they cannot expect corporations to play a patriarchal role in their lives and careers. The Organization Man era, when employees of large corporations could entrust their career to their employer and hope to draw a steady paycheck every month, is drawing to an end.

Even the cornerstone of the Organization Man's existence, lifetime employment, is eroding. As we enter the twenty-first century, professionals around the world are realizing that the concept of lifelong employment is no longer existent. Traditional organizational hierarchies are giving way to project

Box 9.6

MANAGING TECHNOLOGISTS: CONTINUED...

and performance oriented groups and organizational structures, and we are seeing the advent of Gold Collar and Knowledge workers—highly skilled professionals who owe a greater allegiance to their professions than to organizations where they work.

People are starting to take charge of their own careers: professionals are morphing into knowledge workers (a.k.a. gold collar workers) because of the changes in the marketplace. This is especially true for people who started their careers by working for consulting or Information Technology companies, who are acutely aware of the transient nature of their jobs. They are realizing that the best way to remain marketable is to acquire the latest skills and a depth of experience by working on a wide variety of projects and systems, looking at each assignment like a professional basketball or football player looks at his contract with a team.

Motivating *Gold Collar* workers in project teams is an art that Project Managers are already getting better at. Most managers realize that they do not *supervise* or *manage* people but merely facilitate the work done by individuals and teams and ensure that they work towards a common goal of project delivery.

Organizations are poised for international growth and are leveraging offshoring and systems integration and project management. There is a need for special focus on training their workforce, especially the customer-facing, global employees. Apart from the technical skills and knowledge of IT systems, consultants need special focus in the following areas:

- **Core technology skills:** Most technology companies try to equip their employees with knowledge of a variety of technologies and tools. By doing so, they can easily shift people around, based on project contingencies and business needs. Some follow the 'boot camp' approach by putting all new recruits through a rigorous training program and others train employees on a need-only basis. Training and orientation in globalization and cultural awareness is also becoming an area of focus.

- **Team management skills:** IT projects require groups of people to work in tandem, co-ordinating the efforts of team members spread across functional, technical and geographical areas. This is especially true for projects for clients in the onsite or offshore model where a few people working at client sites co-ordinate the efforts of their peers halfway across the globe. Personal issues, differences in personalities and other issues have a way of creeping up if not identified and squashed. All members of the projects need to be trained to handle such issues and communicate fluently and clearly.

- **Project, program management:** Project management is a specialized function in most organizations, with specialists working to co-ordinate projects and deadlines. Members of IT teams need to be aware of the basics of project, program and systems management. Having an overview of the business process being solved and the different pieces that need to fit in order to make a project successful, helps each member of the team to work towards the unified goals. Extending the awareness of program management is the need to be knowledgeable about the intricacies of offshoring workflow.

- **Basics of business:** Most IT initiatives, except for those in the area of R&D, are undertaken for one specific purpose— solving business problems. Naturally, it follows that people working on IT projects need to be aware of the business issues they are trying to solve. This may include functional business expertise like knowledge of Accounting Systems, Financial Systems, Banking, Telecom, Insurance or other areas of business.

- **Communications and cultural sensitivity:** One of the most important aspects of working with people involves communication including oral and written correspondence. This also includes moderating verbal accents and understanding the accents of people from across the globe who speak English in different ways; and may require the assistance of language specialists.

CONCLUSION

In this chapter, we looked at two key aspects of managing a global workforce, viz. cultural, technical and human aspects. Managing global projects draws on the fundamentals and best practices from the management theory and from organizational processes and experiences. During the execution phase of outsourced application development projects, the project manager may not have the luxury or budget to get the entire team together and in some cases, he may not even personally travel to the client's location. Empathy towards members of team from different cultures and proactive use of tools that can facilitate communication are going to be crucial to the success of global development models.

NOTES

1. GaramChai.com: Online portal for the Asian/Indian Diaspora traveling to the west [http://www.GaramChai.com]

2. Indian IT: Time to focus on the positive? [Mohan Babu, *Express Computers' IT People*, 03rd Feb 2003]

3. Correspondence with Benjamin Limbach, PhD researcher

4. Managing cross-cultural issues in global software outsourcing [S. Krishna, Sundeep Sahay, Geoff Walsham; *Communications of the ACM*, Volume 47, Number 4 (2004)]

5. *Global Software Teams, Collaborating across Borders and Time Zones* [Author: Erran Carmel, PHI]

6. The term "B Players" was succinctly articulated in a *Harvard Business Review* article titled "Let's hear it for B Players." [Thomas J. DeLong, Vineeta Vijayaraghavan; HBR Jun 1, 2003]

7. Deena Levine & Associates (DL&A) [http://www.dlevineassoc.com]

8. Frederick Herzberg: Author of several books including *The Motivation to Work and Work and the Nature of Man*. Most famous for the study of what motivates people to work.

9. Abraham H. Maslow [http://www.maslow.com/]

10. *The Cathedral & the Bazaar* [Eric S. Raymond, O'Reilly; Revised edition, 2001]

11. From Organization Man to Free Agent [Mohan Babu, January 2004, *Computer*]

CHAPTER 10

External Landscape and Offshoring Management

- Technology Landscape
- Knowledge Management
- Emerging Techniques from Special Interest Groups
- Globalzation and Economic Environment
- Digital Security and Offshoring
- Staying the Course
- Conclusion

Managing software application development and maintenance from offshore locations to leverage the cost arbitrage is a trend that is expected to gain momentum in the years ahead. In the preceding chapters of this book, we examined several key factors involved in management of IT projects, especially projects outsourced to overseas locations. This is a dynamic field where best-practices, tools and technologies continue to evolve. The Offshoring Management Framework takes a holistic view of offshore outsourcing and is intended to act as a frame of reference that offshoring organizations and vendors can use. In this chapter, we will continue to examine some of the key ideas that individuals, managers and teams can include in their personal and professional roadmaps.

Managers do not operate in isolation: they work closely with clients, business leaders and technologists. They constantly watch for developments taking place outside their industry, and operating environment to scan for events that can impact clients, vendors and other stakeholders. Changes outside the local community, state, or even country can impact businesses in ways unimaginable. In a sense, they have to prepare to ride out *Strategic Inflexion Points*[1], fundamental changes that can affect the operations of their businesses. Sometimes this shift may be due to the change in technology, and at other times, it is because of change in competitive environment, governmental regulation or anything else outside the area of operation of the company. Andy Grove suggests the constant monitoring of corporate radar screens to study changes occurring in the marketplace and preparing for changes when the inflexion point is identified. In a sense, business executives are like controllers sitting in front of radar screens watching movements of each individual speck or dot. Just as a speck on the radar screen can alter the course of other specks, unknown forces can alter the fundamentals of environments in which businesses operate. Examples like the software boom-and-bust cycle are a common occurrence in the business world. (Ref: 10.1: Case in Point).

Offshore outsourcing can induce changes in businesses, the business processes and management of systems. Transition management and change management are evolving into important activities during sourcing. Change management is not a functional area of business per se, but is significant enough that the top management at most corporations are constantly looking out for changes in and around different functional areas. Most major corporations have dedicated groups that study changes happening in and around their area of operation, in the business environment, economy and society in which they operate. Executives and managers also constantly look outside their enterprises to observe the changes in their operating environments since changes in the business environments, government and society have a way of impacting the operations and the bottom-line.

Change management due to offshoring and transition management—taking an organization from an offshoring observer stage to a strong offshoring organization stage—is going to continue to receive a lot of attention. Some of the areas of study and analysis in change management, essentially in the management of offshoring IT systems include:

- Technology Landscape
- Knowledge Management
- Emerging techniques from Special Interest Groups
- Globalization and Economic Environment
- Digital Security and Offshoring

Box 10.1

CASE IN POINT: BOOM AND BUST, AND BOOM

An example of a *Strategic Inflexion point* is the roller-coaster ride of Indian IT professionals. This is an example that's close to my heart since I was also an active participant of the boom and bust cycle of the demand for software professionals.

During mid to late nineties, the global IT industry experienced a sudden boom and the professionals in local markets were unable to meet the demand. Managers of large and small companies lobbied governments for an increase in issuance of visas for foreign workers; among other governments, the US administration decided to increase the quota for skilled professionals from 65,000 to 195,000 per year. Two factors contributed to the boom: the Y2K threat and the dot.com explosion. Most large businesses in America and other western countries with older IT systems began readying them to be compliant for the *ominous* Y2K. Alongside, there was a big IT skill shortage due to the sudden growth in the dot.com and e-commerce sectors

Box 10.1

CASE IN POINT: CONTINUED...

that also required hundreds of thousands of technology professionals. Organizations, afraid of being overtaken by dot.com
startups invested millions of dollars to '*e-enable*' their systems
while at the same time focusing on making their legacy applications to be Y2K compliant. Thousands of Indian IT professionals,
myself included, migrated west to cater to this demand.

The boom did not last long. The technology sector went
bust after the year 2000 when the demand for software
professionals suddenly plummeted after Y2K projects were
completed, and at about the same time, the dot.com bust
dealt a double whammy on the IT Industry. The meltdown in
the US stock market after infamous terrorist attacks of
September 11th 2001 just exacerbated the situation, leading
to a number of large corporate bankruptcies. Hundreds of
thousands of Information Technology professionals were
retrenched by corporations across America. IT professionals
from India in the US found themselves out of job and visa
sponsors, began wondering 'what next'?

Interestingly, this was the time when the 'Global Delivery
Model' began taking off. Executives of Indian IT companies
began to position India as a low-cost outsourcing center and
began to aggressively bid for projects in the west. They also
realized that the thousands of high-tech specialists, who
began to return back to India—armed with experience and
knowledge of the western IT industry—would be a great asset
to their growth plans. Professionals returning to India were
hired by managers eager to build a stronger base. Offshoring
began taking off during this time, a trend that continues even
towards early 2005. Yet another *boom*? Only time will tell.

TECHNOLOGY LANDSCAPE

Offshore outsourcing is one of the *Disruptive Technology* management paradigms, and business leaders are beginning to study the impact of the changes wrought by offshoring on their businesses. Parallels are also being drawn between offshoring and the widespread use of Internet technologies that fundamentally disrupted the way in which we view information. At a time when Internet was being viewed as a disruptive technology trend, a respected strategist, Michael Porter, argued that the Web was just another channel to market goods and services, and shouldn't be considered disruptive. Looking back, Mr. Porter was probably right; however, the widespread use of Internet technologies fundamentally disrupted the way in which we view information. Interestingly, technologies behind Internet (the backbone, browsers etc) lay dormant in universities and research institutions for over two decades before they were "discovered" by commercial users.

Even though Internet and e-commerce is among the oft-quoted 'technological revolutions,' several other innovations in the field of technology continue to occur; and for every innovation that occurs, several fail to materialize. Bridging the '*chasm*' between innovation of technologies and wider general adoption is a term coined by Geffory Moore[2] to articulate this gap. In a sense, Project Managers need to learn to bridge the *chasm* between early adoption of a new technology and the wider general adoption in the market. Changes in technology might not happen overnight but newer application or uses of the same technology can take place almost instantly. All it requires is for one person to think outside-the-box. Similarly, a debate on whether offshore outsourcing is a 'disruptive' trend or just a fad could be just an academic exercise except for the fact that organizations are already beginning to reap benefits.

Technologists are not alone in scanning for emerging trends. Business leaders also constantly look out for tools, techniques and technologies that can give them a competitive advantage. Any 'value addition' that technology or automation can bring to the table can improve the bottom-line. Changes in technology may not occur overnight but newer application or uses of the same technology can take place almost instantly: all it requires is for one person to think outside-the-box. Bhaskar Chakravorti[3], suggests that true insight *"comes from connecting the dots across multiple landscapes (and that) such dots lurk in the unlikeliest corners."* Managers who act as a bridge between business executives, clients and technologist have a ringside view of the landscape, and may need to take on the responsibility of *connecting the dots across multiple landscapes.* For instance, offshoring was a trend that managers and business leaders were observing towards the turn of the century, a trend that has translated to a business practice. On similar lines, another interesting trend in technology management that business leaders are watching closely is the emergence of the open-source software

---

Box 10.2

OPEN SOURCE PROJECT MANAGEMENT: TREND TO WATCH FOR?

The Open Source movement is sweeping through the business-technology landscape in a big way. What began as a renegade movement by a few 'anti-microsoft' activists and Unix enthusiasts to position Linux as an industrial-grade operating system has begun to enter corporate data centers. Corporate leaders and business executives are beginning to examine the business benefits and lowering TCO [Total Cost of Ownership] of open source software and applications.

There continues to be a lot of hype and mysticism around open source, both in the technical and business world. Along with technical journals, the mainstream business media, including Wall Street Journal, Business Week, et al, have been regularly running stories on the open source movement. Although developers have been using open source tools and

Box 10.2

OPEN SOURCE PROJECT MANAGEMENT: CONTINUED...

technologies in projects for a while, selling services bundling open source development is still at a nascent stage. Software service companies are yet to jump onto the open source bandwagon in a big way though many of them are implementing open source projects.

Business leaders and executives are beginning to scan the horizon for trends in open source movement and Managers who can help them cut through the hype are going to have an edge. Managers who are able to guide clients through the SWOT—Strength, Weakness, Opportunity and Threat—analysis and help them prepare a roadmap to embark on an open source movement are going to be in demand.

Will the Open Source movement disrupt the existing application development models? Time will only tell. However, managers need to continue to scan the radar screen for developments on that front.

movement. (Ref: Box 10.2).

Technical evolution is a constant process and the spoils always go to business leaders and technocrats who can identify and commercialize ideas. Any 'value addition' that technology or automation can bring to the table can improve the bottomline, increasing competitive advantage. Business leaders need to be on a constant lookout for technologies, tools, and techniques that can give them a competitive advantage. For instance managers in the retail industry constantly lookout for advancements in the supply-chain management sector that can help them manage their inventories and reduce costs. As the trend towards offshoring continues to gain momentum, newer models of managing sourced work, and tools and technologies to facilitate offshore-onsite coordination are going to emerge. Managers need to continue to scan the landscape for such emerging best-practices and leverage them.

KNOWLEDGE MANAGEMENT

Management of organizational knowledge, learning's and history is receiving increased attention from senior executives as offshoring projects take off and Managers leverage the collective learning and experiences of groups and teams to funnel back to the organizational knowledgebase by applying some of the practices of Knowledge Management (KM). By using KM techniques and technologies, managers can leverage the intellectual capital of their organizations by helping knowledge workers optimize their skills and potential. Offshoring vendors, along with consulting organizations continue to make strides in adopting such systems since they understand their own 'requirements' and are best positioned to deploy and use these techniques; eating one's own cooking, if you will. Managers are trying to leverage the intellectual capital of organizations in a number of ways:

- **Use of Knowledge Systems and software:** Enterprises are constantly attempting to link most of their software systems and provide their knowledge workers with a unified view of the organization. Knowledge portals and Intranets are increasingly being used for this purpose, and are helping in faster and more accurate decision-making. Tremendous amount of research is going on in the field of knowledge servers; multi-use ontologies, taxonomies, knowledge bases and knowledge system modules. Consulting companies have been at the forefront of applying knowledge sharing systems. They realize the value of their key resource—the intellectual capital of their employees lies in the collective growth. Many of them have not only built huge databases and FAQ (Frequently Asked Questions) systems but have also been working on motivating their employees to update and maintain those databases. The knowledge databases can be as simple as a list of questions with answers, but are generally sophisticated enough to help users with advanced querying.[Refer: 10.3—Knowledge Management at work]

- *Knowledge of Technologies:* Experience and knowledge of the best practices of application development helps managers appreciate the nuances and issues faced by their team members. While it is not mandatory, technical knowledge can be a valuable asset, especially as managers also have to co-ordinate with peers, customers and executives and it is imperative that the technicalities and details are not lost during such communication.
- *Knowledge of domains:* Knowledge of the business domain is a key success factor for project managers. While they will generally depend on business analysts and domain experts for guidance, any knowledge of the functional area will go a long way. Because of the complexity of businesses systems, it is hard to expect an individual to gain expertise in all the different functional areas in a reasonable amount of time. Well designed Knowledge Management systems can help managers capture and retain domain knowledge for subsequent projects and programs.

- **Knowledge culture and Business Systems:** Knowledge Management systems cannot operate without sufficient mature knowledge sharing culture and practices, including:

  - *Knowledge sharing:* Building large knowledge sharing systems is the first step toward intellectual capital management. Companies are beginning to realize that real knowledge sharing comes when there is a culture of sharing and trust in the organization.

- *Knowledge transfer:* Unlike other products and commodities, knowledge enhances in value by constant use and sharing. Knowledge transfer involves transmission of knowledge behind use of technologies, best practices and techniques across teams, groups and divisions in the organization. By fostering a culture of sharing best practices and ideas, knowledge can be transferred seamlessly across the organization.

- *Experiential learning:* People acquire new skills in two main ways: cognitive and experiential. Cognitive learning corresponds to academic knowledge such as learning vocabulary or multiplication tables, and experiential learning refers to applied knowledge such as learning to mow the lawn or to drive a car. Knowledge can also be acquired by experience and hands-on learning. Many large companies have formal on-the-job training programs and orientations that most new employees undergo. Undergoing a formal experiential learning program helps trainees solidify the knowledge acquired in classrooms and teaches them to apply their skills to solve real-world problems.

Successful organizations and managers build a culture where knowledge sharing and transfer is actively encouraged and rewarded. They use a variety of mechanisms including seminars, written memos, internet newsgroups, mailing lists, newsletters, oral presentations, online multimedia demonstrations, site visits, job rotation, education and training to encourage knowledge transfer. Many companies also have an ongoing practice of project success and failure analysis and review. This helps a larger number of employees learn from the experience of a few.

Box 10.3

CASE IN POINT: KNOWLEDGE MANAGEMENT AT WORK

In my previous job, I consulted as a Project Manager for a large network-engineering group of a Fortune 500 Telco in the US. The developers in our group would work on various aspects of the system including building new enhancements, testing/troubleshooting and maintaining the existing systems. In our group, we had a person—let's call him Mr. X—who was dedicated to handling tech support calls and trouble tickets, a job he had been doing for a number of years. The first line of technical support in our organization was handled by the call-center representatives, who would direct only the most complex calls to Mr. X. He was the "go to" guy in our group, whom even our seasoned analysts and developers went to. He had a ringside view of our system, bits and pieces of which he would look at on a day-to-day basis.

One morning, with no advance notice, Mr. X had to be let go because of "budget cuts". The senior management and the team were frantic since everyone feared that the organizational knowledge had walked out of the door with Mr. X. I was asked to go scrambling to find his replacement.

In the back of my mind, however, I was not as perturbed as others. Within a couple of days we hired a replacement for Mr. X, who was able to firmly hit the ground running, with very little lead time or learning curve. People began wondering how it was possible. My team and I had planned for such contingency by designing a "database tool" that also happened to double as a Knowledge Management system.

When we built the "database tool" a few years ago, none of us knew anything about Knowledge Management. I don't think KM was even a buzzword then. The tool was quite straightforward, built using IBM's Lotus Notes. It had forms for updating all the trouble tickets that we encountered along with a brief description of the resolution of the problem. The

Box 10.3

CASE IN POINT: CONTINUED...

tool also had a MIS element to aid in reporting the most frequent problems, statistics etc. Lotus Notes also has a built-in sophisticated search mechanism that we were able to use in order to search for 'similar problems' using English or technical keywords. Of course, after the system had been built and deployed, management had the foresight to decree that anyone working on trouble tickets had to update the database too. This was useful when Mr. X went on vacations or was sick and proved invaluable later.

When Mr. X suddenly left the organization, he left behind a small, albeit, significant part of his knowledge in the form of this knowledge-base. I didn't know it then, but I was managing a Knowledge Management system that had paid for itself manifold.

Case reference: Author's column[4]

The success of outsourcing hinges on successful knowledge acquisition, more so in an offshoring context where information and knowledge of systems gathered in one location will have to be transmitted for use and implementation elsewhere, probably in a location separated by time-and-space, and cultural boundaries. Knowledge management in an offshoring context is acquiring greater scrutiny among senior management and business leaders in another context: retention of in-house knowledgebase when systems are sent offshore. This issue of knowledge retention when systems are outsourced had been addressed by systems integrators and software vendors in the past; however, outsourcing IT systems to offshore locations is bringing renewed focus on knowledge retention strategies. Tools and techniques of Knowledge Management are increasingly being used to assuage concerns of offshoring organizations and executives. The nascent area of managing knowledge in distributed projects is also leading to a lot of research; as an example, the authors[5] say *'Knowledge in projects calls for a close look at insights generated within each individual project, such as schedules, milestones, meeting minutes, and training manuals. Individual project members need to know when, what, how, where, and why something is being done and by whom, with the goal being to promote efficient and effective coordination of activities.'*

EMERGING TECHNIQUES FROM SPECIAL INTEREST GROUPS

Professionals and Managers regularly participate in special interest groups outside their organization to improve their understanding of the business landscape. Membership in industry and special interest groups along with regular attendance in peer-groups meetings and industry gatherings also help professionals learn new concepts and keep abreast of latest happenings in their fields apart from helping personal and professional networking. Such groups and bodies also help in formulating 'Position Statements' on emerging trends including offshoring. Refer to Appendix A for brief discussion on some industry bodies that are active in the global arena along with a sampling of some position statements from professional bodies on offshoring.

Special interest groups are not restricted to professional associations alone. They include consumer user groups, labor groups, chambers of commerce and other professional bodies can lobby for and influence the course of their business and services. Such groups serve an important function: they ensure that decision makers in both the government and corporate world hear the voice of the professional community. For example, offshoring endeavors also lead to some relocation of jobs and shift in careers in the communities where the organizations are based. Such shifts may lead to resistance to offshoring within the organizations and in the society. Managers should work towards addressing such concerns while ensuring the success of offshoring initiatives by striking a delicate balance. Executives and managers can ideate and seek the assistance of industry bodies and groups in consulting and formulating strategies to mitigate societal concerns.

Technology vendors[6] have also created sub-cultures around special interest groups [also called user groups] among members of the technical community. They anchor regular meetings, seminars and events to evangelize their technologies and newer innovations along with creating specialized portals and internet forums. Though the intent of such groups is to promote their technologies and market their software, such groups can provide managers insight into emerging trends.

Participating in professional groups and being aware of the activities of Special Interest Groups in the community and industry can help executives, technologists and managers to keep abreast of the latest trends; and some of the benefits of participating in such groups include:

- Lobbying for common interests and goals
- Influence governmental and societal policies
- Influence adoption of technical and industry standards
- Professional Networking
- Enhance the professional Body of Knowledge

Executives and managers may not always be able to participate in the activities of all the key groups and bodies; however, they should attempt to keep abreast of trends emerging from such groups that may benefit them or influence the course of their organizations. Lobbying with industry bodies in forums and working with government policy makers is increasingly receiving focus among managers, especially as pockets of resistance to offshoring begin to get mention in the media. Managing such external expectations and trends is among the emerging challenges of technology and business leaders. An understanding of the intricacies of micro and macro-economics can help business leaders better understand the social, political and business environments in which they live and work.

GLOBALIZATION AND ECONOMIC ENVIRONMENT

Businesses rarely operate in an isolated cocoon; there are always a multitude of stakeholders vying for the attention of the top management. Business partners, vendors, suppliers, clients, employees, shareholders and investors, governmental institutions are all typical stakeholders of any organization. Business leaders constantly examine the impact of economic conditions locally, domestically and internationally and strategize on how such changes in the economy affect the operations of their organizations. Scanning the economic landscape includes an understanding of the financial, social and political context in which the economy operates. Changes in the marketplace could include a combination of governmental, economic or technological changes. Sometimes, the barriers to entry in the marketplace may be reduced or vanish because of a combination of some of the factors or by the entry of a few powerful rivals. Many such indirect changes may not be obvious but experienced managers and executives depend on their 'gut instinct' to gauge the depth of such changes and deuce cause-and-effect relations, especially on their industry or business. Changing economic climates also influence profitability, billing rates and operating margins which managers need to watch for.

Service delivery organizations are generally tuned to trends in the marketplace since it can impact the demand for their services. An example of this is the increasing surge in globalization of Application Development and Maintenance efforts, initially driven by companies looking to cut costs while continuing operations. Software Service firms that had already evolved their offshoring methodologies have begun to expand their market footprint.

Managers, especially those managing global delivery initiatives need to be acutely tuned to economic and geopolitical trends. Awareness of such trends will help them empathize with the changing needs of clients and help them plan for solutions and offerings tuned to the needs of markets. Such empathy may sometimes translate to newer business opportunities. George A. Steiner, *et al*[7], in their book look at different models conceptualizing the relationship between businesses and the government and society. The two models that stand out are the dominance model and the stakeholder model. In the dominance model, the government and businesses dominate the great mass of people in the society. This model of business, governmental and societal interaction is also a reality in many developed nations like Japan, Korea and China. For instance, in China the family relationships between business leaders help foster trust and an unwritten code-of-conduct. In the stakeholder model, the business firm is at the center of a set of mutual relationships with individuals and groups called stakeholders. Even in free economies like the US, the Government still wields tremendous sway over commercial and fiscal affairs of its citizen. For instance, even to download a copy of a commercial software or security patches from Oracle or Microsoft, one has to agree to a declaration like *"Eligibility Export Restrictions"* and certify that *"I am not a citizen, national or resident of, and am not under the control of, the government of: Cuba, Iran, Sudan, Iraq, Libya, North Korea, Syria, nor any other country to which the United States has prohibited export....."*

Changes in policy or government can also affect businesses policies and strategies, a fact that managers and business leaders need to constantly keep an eye out for. While stakeholders may not influence individual projects, the operations of businesses could be influenced by them. Some of the aspects of policies that could impact business include:

- Changing Governmental regulations and restrictions including industry specific regulations
- Impact of specific regulations on conducting business both domestically and internationally

- Fiscal and monetary policies including tax structures and incentives from governments could sway the offshoring and other sourcing strategies
- An understanding of the levels of bureaucracy and red-tapism; this is more significant in a multinational context where the policies and bureaucracies of multiple nations could come into play

A Case-in-point of governmental policies directly impacting offshoring: The work-visa regulations are highly nebulous and can impact the implementation of the Offshoring Management Framework since project team members can be constrained from traveling. Offshoring projects could be influenced by governmental regulation and restrictions on travel including the H1-B work visa policy of the American government; and other such policies from western nations. Executives need to be updated on travel advisories, visa restrictions and changes that could impact offshoring initiatives. Other aspects of external environment impacting offshoring include the need for executives to lobby against anti-sourcing legislation and managing public relations to sway public opinion in a global environment.

DIGITAL SECURITY AND OFFSHORING

Offshore outsourcing is drawing renewed attention on Information and digital security in the corporate world. At the systems side, all major software vendors—including Microsoft, IBM, Oracle, *et al*—have promised to attack the threat at its roots by making the software and operating systems more impregnable, but executives continue to higher assurance from application developers and IT managers. The recent spate of *virus* and *worm* attacks have brought home the significance of informational security, crippling computer systems and communication across the globe, causing billions of dollars in losses. Business magazines and management journals continue to run detailed features on virus attacks and security, a testimony to the seriousness with which managements are looking at this issue. A succinct Harvard Business Review article[8], talks about three main categories of threats to digital security:

- **Network attacks:** Waged over the Internet, network attacks can slow network performance, degrade e-mail and other online services, and cause millions of dollars in damages. Most new enterprise security tools can thwart common network attacks, and even if the systems are knocked out, the damage is rarely permanent.
- **Intrusions:** Intrusions differ from network attacks because the intruders actually penetrate an organisation's internal IT systems. Hackers use human skills, also known in the industry as "social engineering," to get insiders in an organisation to spill out vital details about the security. Once inside the system, intruders masquerade as genuine users to create havoc.
- **Malicious code:** Malicious code consists of viruses and worms which can wreck havoc faster than human hackers. Viruses need help replicating and propagating, whereas worms do it automatically. Because they are propagating themselves through the systems and networks, their targets can be random, making it impossible to predict where they'll hit next.

Software companies regularly play 'cops and robbers' with hackers. The cops include tools like the virus guards in their arsenals and are employed by security software companies and information security departments of organizations. Information security management is a niche area of IT that has gained prominence in recent years and many companies are investing heavily to protect their systems. Manoj Kumar[12], an Information Security expert with a multinational says "*Business leaders and IT managers should work on a customized solution for each aspect of their organization's systems. The level of security should depend on the nature of the asset. For instance, a financial institution will have to focus on building a high level of security for its core financial application and interfaces with other organizations and the Web portal. If the resources are limited, it may do so at the cost of providing extra-high security for its informational portal"* and goes on to add *"The analogy here is to a supermarket which locks its expensive perfumes, jewellery, and CD players behind a glass door; whereas it may only have a few security cameras around its grocery or produce section.'* Information security and risks are also gaining attention in the offshoring context as evident from the viewpoint by Nancy Mead[10] who has articulated some of the important trends in this sector.

Box 10.4

VIEWPOINT: OUTSOURCING AND INFORMATION SECURITY RISKS

When outsourcing is the subject of discussion, issues that surface typically include concerns about employment for U.S. engineers [1], the lower salaries in developing countries, the adequacy of management, and communications problems. These business issues are indeed important, but little attention is being given to the potential for information security problems when outsourcing occurs. Some of the issues are:

- How to determine whether the software developed, maintained, or enhanced offshore is trustworthy
- Whether certification of the developers or the companies is an appropriate method for assessing trust
- Types of software that should or should not be outsourced. For example, should software that will be used in critical infrastructure or the financial markets ever be outsourced?
- Alternatively, consider widely used COTS software, such as Windows. If COTS development is outsourced, could a time

Box 10.4

VIEWPOINT: CONTINUED...

bomb be embedded in the software that would avoid detection and cause many systems worldwide to crash?

- Concerns with assuring the privacy of data related to outsourced applications [http://www.crmbuyer.com/story/California-Legislation-To-Protect-Consumer-Data-Overseas-36136.html]

It's fairly well known that the number of reported information security incidents reported to CERT **http://www.cert.org** has grown roughly exponentially from one year to the next. Established in 1988, the CERT® Coordination Center (CERT/CC) is a center of Internet security expertise, located at the *Software Engineering Institute*. This exponential growth of incidents parallels the growth of the size of software (lines of code or function points) in use. So it may be that on the whole our ability to engineer software without vulnerabilities has really not improved. There are other possibilities as well. It may be that the number and variety of attack methods has increased as well. Furthermore, as attack sophistication increases, intruder knowledge decreases, in the sense that intruders do not need to be as knowledgeable in order to launch an attack. This occurs because hackers are able to easily pass around attack scripts. Another source suggests that in a computer crime survey, some 90% of respondents detected computer security breaches of one form or another [2].

Against this backdrop we have started to see attention to information security, with major vendors such as Microsoft making security a priority item. The now famous Gates memo resulted in some significant changes within Microsoft, as we see in a retrospective discussion a year later by Michael Howard, a Senior Security Program Manager at Microsoft, and Steve Lipner, Director of Security Assurance at Microsoft [3]. For one thing, during the months of February and March

Box 10.4

VIEWPOINT: CONTINUED...

2002, all Windows feature development stopped, while the design, code, test plans, and documentation were analyzed by the Microsoft team. For a company that is market-driven to provide more and better features, this is an unusual step, and indicates just how seriously this security initiative was. Training courses were developed and delivered to support the "Windows Security Push." As part of this effort, it was observed that security is not just a "layer" that is added after the fact, but a consideration that pervades all of development. In Howard and Lipner's words, "Secure software means paying attention to detail, understanding threats, building extra defensive layers, reducing attack surface, and using security defaults." This is a daunting task, and even if all the new code can be made secure, what about the old code that remains embedded in these applications? The amount of legacy code in Microsoft products suggests that this is not an easy problem to solve.

One article suggests that the most damaging and difficult to detect malware is that which is designed to attack a specific enemy [4]. An example of such an attack is malicious code inserted by a disgruntled employee or a sophisticated attacker such as a moral activist. Many times the measures taken to protect against insider threat are improved processes, in addition to technical solutions such as encryption, signature keys, etc.

These considerations lead directly to the question of whether outsourced software can be trusted.

There are many opportunities to insert malicious code in software. This could occur during the transfer of the software from the subcontractor to the contracting organization. It could occur as a result of a successful attack on the subcontractor's system(s). Or it could occur if one of the subcontractor's

Box 10.4

VIEWPOINT: CONTINUED...

employees has malicious intent. In spite of international cyber-crime agreements, it appears that laws are enforced to varying degrees in different countries [http://www.mercurynews.com/mld/mercurynews/business/technology/9500402.htm?1c]. Of course, cybercrime can occur anywhere, but we need to consider the following special situations when outsourcing:

- a cyber attack by terrorists whose aim is to disrupt critical systems
- a financial attack by gangsters who wish to transfer funds to their own accounts
- attempts to set up software to launch targeted distributed denial of service (DDOS) attacks from large numbers of systems
- theft of intellectual property (industrial espionage)

Some mechanisms to improve trustworthiness include:

- use of encryption to protect transmitted code
- use of analysis techniques such as function extraction to detect the insertion of malicious code
- establishment of trust mechanisms between the contracting organization and the subcontractor

There are many other candidate mechanisms for addressing the problem. However, until there is sufficient awareness and concern with the information security risks associated with outsourcing, this will continue to be a problem area. In the meantime, executives who are considering outsourcing should use the mechanisms that are available to them today: Assessments (a la CMM) of the organization's processes, background checks on employees, risk assessments, review of physical security, and review of political situations in the country that could result in higher risk. In addition, good old-fashioned management practices can go a long way. If you

Box 10.4

VIEWPOINT: CONTINUED...

were developing software inhouse or with a local contractor you would want to track it pretty closely. The same tracking mechanisms need to be applied to outsourced software projects.

1. Costlow, T. "Resources: Tough in the Trenches." *IEEE Spectrum*, Vol. 41, No. 7, July 2004, pp. 51–54.
2. Rabinovitch, E. "Securing Information Technology Infrastructures." In *Proceedings of the International Conference on Communication Technology (ICCT)*. IEEE Computer Society Press, 2003.
3. M. Howard and S. Lipner, "Inside the Windows security push," *IEEE Security & Privacy*, vol. 1, no. 1, pp. 57–61, Jan.–Feb. 2003.
4. Wright, A. "*Controlling Risks of E-Commerce Content.*" Computers & Security, Vol. 20, No. 2, April 2001, pp. 147–154.

STAYING THE COURSE

The organizational strategies geared towards offshoring may generally follow a *top-down* approach where program management strategies for offshoring multiple projects, programs and initiatives are formulated for the organization and individual projects and programs rolled out. Setting the course for offshoring, transitioning to a steady state and reaping promised benefits continues to be a key goal. At the lowest level of the offshoring granularity of individual projects, Managers are tasked with ensuring that the onsite-offshore coordination functions proceed smoothly as expected. Monitoring of project activities pertaining to a Project Life Cycle commence right after the first project task begins and continues till the project deliverables are successfully handed over.

Perhaps the most important responsibility of an offshoring management team is to ensure that projects and programs stay on course. This is significant since most projects rarely ever run according to plan; offshoring may add newer challenges that may have to be addressed along with regular managerial challenges. Various parallels to this activity of ensuring that the project stays on track with other real-time management activities abound, including my favorite, the 'airplane analogy.' In his book[11], Ed Sullivan draws an interesting parallel between the *flight plan*, *unpredictability of the trip*, *navigation system* and the *descent* of even a 'typical' flight of an aircraft flying from Boston to San Francisco. He says, '*During the flight, there will be countless factors that could potentially push the flight off schedule. Similarly, projects grapple with issues pertaining to budgets, resources, people and slippage of schedules and deadlines. Managers, like pilots, need to be extra vigilant in ensuring that they have a system in place to constantly monitor progress.*' By efficiently monitoring the landscape and work product outputs of a team, a manager can initiate preventative action on tasks and activities that may slip in schedule.' Experienced managers learn the art of peering at an invisible 'radar screen' to find their bearing without having to double-check. Aided by tools and techniques including project tracking and monitoring software, they can uncannily zero in on bigger problems while letting their team grapple with routine issues. [Refer: Box 10.5] In an offshoring context, some of the external landscape issues that managers will have to add onto their radar screens are the art of evaluating the shifting Technology Landscape, applying some of the emerging techniques of Knowledge Management, trends from Special Interest Groups and Economic Environment.

Newer tools and techniques for measuring and monitoring projects and tracking progress continue to emerge. There are few tools to manage offshoring, most of them developed by offshoring vendors for their internal use. However, automated status gathering includes use of web based alerts, collaborative tools and Project

Box 10.5

CASE IN POINT: A SENSE OF WHERE YOU ARE?

My boss, a veteran with a few decades in the industry under his belt, is passionate about improvements in the field of project monitoring and tracking. One of his favorite books, going back to the days when he was in the national basketball team is Pulitzer prize winner John McPhee's story on Bill Bradley[12]. Like McPhee, he is fascinated by Bradley's uncanny ability of always knowing his position in relation to the basket while playing in the court. This ability to intuitively figure out one's position during the game also triggered the title of the book. Though published a few decades ago—much before its relevance to the current IT generation—the central idea perhaps holds true even today.

Those who have played or watched basketball, or for that matter any contact sport, can perhaps empathize with the need to know one's precise position during the thick of a game. Really good players hone the art of knowing one's orientation without having to double-check. This ability to know one's bearing without having to cross check can spell the difference between success and failure in a fast game. This includes the ability to know where one's peers and partners are placed in the court since one may have to pass the ball to the person most able to take a successful shot so that the team can win.

Managers can sometimes get bogged down in the details of planning and operations and may find that they are loosing sight of the big picture. Knowing the pulse of the project, how members of the team are faring and having a handle on all the issues is an art managers begin to acquire as they manage successive projects. An argument can be made that the concept of *'A Sense of Where You Are'* can be extended to project management. Managing on the fly is an interest concept that other management authors and thinkers have explored, perhaps using other analogies.

Management tools such as Microsoft's Project[13] among others are increasingly being adopted to manage offshoring projects as work-flow and reporting in an offshoring context is assuming greater significance. Other techniques including Business Intelligence, project dashboards and project health trackers are also becoming popular in this space and are especially useful for tracking the 'health' of large and complex projects. The caveat is that even such tools and dashboards are only as accurate as the inputs that feed in; as the GIGO maxim goes: Garbage In, Garbage Out.

Offshoring or the transition management and workflow from one culture and geography could also induce changes that have to be planned and managed. Every offshoring project, program or initiative induces changes in the operations management including changes in the activities of the life cycle. People and processes need to adapt to changes induced by transformation. They may need to learn newer communication techniques, tracking workflow and working with peers to ensure that the IT systems continue to support the need of business users. Existing change control policies, tracing and management of change-requests and work-requests and gathering requirements for application enhancements and development of new systems may also have to be tailored to manage offshore projects. The transition of an IT system from the offshoring observer stage to a strong offshoring organization stage (as described in Chapter 3) will include strong program and change management activities. Managers will have to ensure that their organizations continue to stay the course while also ensuring a steady transition.

CONCLUSION

Offshoring, including managing people from across cultural and geographic divides is among the emerging focus areas. Aspects of external landscape including changes in technologies, lobbying by special interest groups, economic upheavals or governmental policies can directly or indirectly influence projects. The fast technological innovations may sometimes lead to a slow pace of adoption, and managers may be expected to bridge this chasm between early adopters of technology and the wider market adoption.

Globalization and delivery of projects across geographic boundaries is beginning to take off and the proposed Offshoring Management Framework and is only bound to be revised based on emerging practices from the field. In this book, we examined a few emerging ideas and trends in Offshoring Management and looked at how managers and executives can benefit from borrowing from some of the best practices in the industry. The ideas presented are by no means exhaustive, and organizations, managers and academics continue to add to the body of knowledge.

NOTES

1. Andy Grove coined the term Strategic Inflection Points in his book [Only the Paranoid Survive : How to Exploit the Crisis Points That Challenge Every Company, Andrew S. Grove]

2. Crossing the Chasm: Marketing and Selling High-Tech Products to Mainstream Customers [Geoffrey A. Moore, HarperBusiness; Rev edition (July 1, 1999)]

3. The Slow Pace of Fast Change: Bringing Innovations to Market in a Connected World [Bhaskar Chakravorti, Harvard Business School Press (June 12, 2003)]

4. Knowledge Management, still a buzzword? [Express Computers' IT People, 9th Apr 2001]

5. Managing Knowledge In Distributed Projects [By Kevin C. Desouza and J. Roberto Evaristo, April 2004 Communications Of the ACM]

6. Group on the web listing some of the popular technology user groups [Top: Computers: Organizations: User Groups] http://www.supercrawler.com/Computers/Organizations/ User_Groups/

7. Business Government and Society A Managerial Perspective [George A. Steiner, John F. Steiner; Irwin/McGraw-Hill]

8. The Myth of Secure Computing [Robert D. Austin, Christopher A.R. Darby; Harvard Business Review, June 2003]

9. Reference: Interview with Manoj Kumar, an Information Security expert based in Bangalore, India

10. Nancy R. Mead, PhD from the Software Engineering Institute [Carnegie Mellon University] wrote an article on 'Outsourcing and Information Security: What are the risks' for Cutter IT Journal [October 2004, Volume 17, No. 10] and provided an abstract viewpoint for this book.

11. Under Pressure and On Time [Ed Sullivan, Microsoft Press, April 4, 2001]

12. A Sense of Where You Are : Bill Bradley at Princeton [John McPhee, First published in 1965]

13. Microsoft's Project [http://www.microsoft.com]

CHAPTER 11

An Offshoring viewpoint on Pre-sales

- Overview of Pre Sales
- People Supporting Pre Sales
- Responding to Pre-Sales Requests
- Best Practices in Information Technology Pre Sales

Pre Sales includes the entire gamut of activities involved in preparing to engage with prospects, clients and others and includes specific responses to client requests. Clients or companies that need software services and project implementations generally call for proposals (RFPs) where they expect formal responses from their vendors and service providers. Although it is hard to generalize on the nature of or the contents of such proposals, most documents follow a structured framework: detailing the project, asking vendors for suggestions or solutions or proposals along with cost estimates regarding the work to be done.

OVERVIEW OF PRE SALES

Typical Pre-sales support activities include:

- **Responding to client requests:** Responses to clients could include informal responses, pointers to publications, colleterals or other references or take more specific forms like responses to proposals including: Request for Proposal (RFPs), Request for Information (RFI) and specific Statement of Work (SoW) or Work Orders

- **Supporting client visits:** In some cases, clients or prospective clients may make a trip to offshore vendor's offices for a personal visit prior to engaging with them. This could include offshore client visits targeted at offshoring

- **Visiting clients and/or making presentations:** Engaging clients for larger, complex deals involves a number of activities, including making presentations, meeting with clients to discuss specific aspects of their (client's) initiatives, to get a better understanding of the context in order to make specific recommendations in proposals. This may also include preparing proof-of-concept demonstrations and solution mockups.

- **Competitor Analysis and market scanning:** This is a crucial aspect of pre-sales since many clients evaluate responses from multiple vendors, and responses should address such competitive scan. The analysis could include using online tools, subscribing and analyzing research reports, analyst studies, market research data etc.

- **Sales Support:** Such activities may include supporting sales and account teams in responding to general client queries about solutions and capabilities. This could include partnering with onsite/client facing Sales or Business Development Managers to identify and convert prospects into customers.

- **Interfacing with other internal groups** (within the organization) while responding to client requests. This is especially true of larger software service firms where Pre-sales people from one group/division may have to rope in Subject Matter Experts from other groups while responding to a client request or proposal

- **Marketing support:** Large service firms work hard at differentiating themselves from others by formulating marketing messages and evolving Go-to-market solutions or customized offerings. This may also take a form of alliances with other software product development firms or niche vendors. Pre-sales activities may include leveraging such alliances to showcase extended capabilities to clients.

PEOPLE SUPPORTING PRE SALES

A vast majority of technologists work for consulting or software services companies, while a smaller percentage works for end clients. This is partly attributable to the trend towards offshoring and outsorurcing. Increasingly, service firms are also dedicating professionals to work on pre-sales and sales support activities on a full-time basis. Many of them come from marketing or sales background and follow a well-defined operating process involving plugging the response documents with common templates about the company and its capabilities. However, the marketing people may not have the same depth of experience in technology to respond to all aspects of RFPs. For instance, skills required to customize solutions to project and client specific responses, which may require someone with a technical background. Technical people will be able to analyze the client's problem, and think through a framework to create a solution based upon their knowledge and experience.

The full-time staff responding to RFPs may be augmented by those on un-billed (bench) time. While preparing project proposals, and responding to Request for Proposals (RFPs), consultants gain expertise in the 'business' of software services while helping address the "how to" part of the problems.

Even though technical people can be of invaluable assistance to pre-sales project teams, most techies are loath to be involved in such work. There are a number of reasons why techies abstain from being involved in pre-sales support work:

- Sales support is a repetitive work: Most responses to RFPs involve "cut and paste" from seed documents and various sources—a task which technocrats find monotonous.

- Lack of instant gratification: Pre-sales cycles are generally long, and it takes weeks (or months) before the results of a proposal can be known. This is the reason pre-sales people work on multiple proposals at any given time. Techies, on the other hand, come from a background where they can "see" the results of their code or work almost instantly.

Larger companies, especially the 'big five', weave incentive plans, bonuses and career growth around such "corporate activities," typically expecting consultants to log 15 percent to 20 percent extra time on such initiatives. Using intranets, VPNs, remote logins, and sophisticated workflow tools, companies are able to track the activities of consultants to reward and motivate them. Many have tried building large knowledge management systems by adding a repository of frequently asked questions, how-to's, past projects, case studies, standardized response templates, etc.

RESPONDING TO PRE-SALES REQUESTS

Clients or companies that need software services and project implementations generally call for proposals from a pool of preferred vendors. Although it is hard to generalize on the nature of or the contents of such proposals, most documents follow a structured framework: detailing the project, asking vendors for suggestions or solutions or proposals along with cost estimates regarding the work to be done. RFP responses would generally involve two components:

a) The "How To" part that addresses key questions on how the service provider will help the client organization find optimal solutions. A typical response to an RFP or proposal will include a substantial technical component. People responding to RFPs at service firms generally follow a well-defined operating process involving plugging the response documents with common templates about the company and its capabilities. The customization process kicks in when it comes to project and client specific responses; and here is where someone with a technical background is really valuable. Technical subject matter experts are needed to analyze the client's problem, think through a framework to create a solution based upon their knowledge and experience. Such skills can be especially useful while preparing a proof of concept or technical demo. The focus areas include:

- A technical solution architecture, approach or framework to solve the problem. The intent is to demonstrate to the client that you Get their problem and showcase how you will approach the solution. During Pre-sales phase, technical solutions could include a mockup of the end-state technical view, reference architecture, approach or framework to solve the client's specific problem.

- Case studies, proof of concept, demonstration or mockup to showcase how the service provider has solved similar problems elsewhere. Organizations typically demonstrate their capabilities by referencing past successes (Case studies, whitepapers etc), and may also develop proof-of-concept (POC), demonstrations or mockups.

b) The "why" component which includes details business leaders and "management" would be interested in. Commercial and administrative aspects include a whole gamut of activities involved in responding to clients with specific reference to the processes involved in executing the engagement / project. Cost is definitely a key criteria organizations use while evaluating a proposal though depending on the nature of problem being sourced, the credentials of the vendor and the solution may take a higher priority. The administrative aspects include a high-level estimate of the effort involved in terms of duration (time), effort (people/resources) and additional resources including infrastructure etc required to successfully provide the required solution. Estimating the level-of-work involved may include formal estimation techniques based on expertise from past projects or could be a very heuristic process, especially for newer technologies without adequate benchmarks. The focus areas may include:

- Cost, budget and financials: What is the total cost to the client, how often will they be invoiced and the mode of payment etc? This may include defining the billing model: Time and Material (T&M), Fixed Price (FP) or other blended models.
- Staffing plan, resource management: Responses to proposals typically include staffing plans (how many people, skills they bring to the table, roles etc) and may also include other resources needed including specific systems, hardware, software etc.
- Credentials, testimonials and references from past clients. There are instances where clients may ask for

specific testimonials from existing/past clients of service firms. Staff engaged in pre-sales activities should be able to arrange for such references.

BEST PRACTICES IN INFORMATION TECHNOLOGY PRE SALES

Perhaps the most important tools to consider are the organizational resources. It is easy to forget that while responding to client queries in a pre-sales role, you are not an individual but an ambassador of your company. Large services firms have several tools including

1. An elaborate Knowledge Management portal accessible to employees across the globe. The taxonomy is intuitive and searchable. Typically people responding to proposals scan the KM portal for internally published data and references on projects, technologies

2. Email groups: There are several e-mail groups of experts (functional, technical and others). Responses are near-real-time as they can be!

3. Bank on researchers and internal experts: There are bound to be several research groups – technical, functional and business focused – whose focus is on helping client facing teams address client issues and challenges including responding to pre-sales queries. You will need to find your internal research groups in your organization that you can ping.

4. External research: this will help in landscape scanning, getting information on competitors, market trends and the like. This could be formal or informal research (depending on the nature of the query, cost, budget and time required to respond)

I realize that some of the suggestions above are skewed towards larger organizations that have invested in research and support for pre-sales and customer servicing. However, managers at smaller organizations also have similar, though informal, researchers and 'thought leaders' they can bank on. You will have to network with peers in your organization to discover the support mechanism available to you.

Perhaps one of the most important criteria I have personally observed to be helpful is the RTQ and ATQ principle:

- RTQ = Read the Question. Refer to what the client is asking for in the proposal

- ATQ = Answer the Query: After you have read and understood the question (and researched on it) you should focus on just Answering the Query. And remember: say "NO" "NA" or "info not available" if you don't have the answer. Saying so will help you more than hurt you. Nobody (least of all your client) wants to spend precious time reading an answer to a query they did not ask!

NOTES

Readers may notice that this chapter is skewed towards those from software service companies looking to respond to RFP's. this is intentional since responding to RFP's is a complex craft which is at a nascent stage in development. Evaluating RFP's and supplier management, on the other hand is much more mature in organizations. Technologists at the other end of the proposal management, at the client organization will get involved in reviewing proposals and responses from vendors. The best practices in vendor management and evaluation for non-offshored outsourced projects can and should be leveraged even while evaluating offshored projects.

CHAPTER 12

Enterprise Architects Enabling Strategic Global Sourcing

- ☐ Enterprise Architects and Sourcing
- ☐ Managing and Addressing EA Challenges in a Sourcing Context" (Capitalize EA)
- ☐ Loss of Technical Expertise Due to Sourcing

Recession? TCS, Infosys, Wipro, HCL Bag Large Outsourcing Deals.
— **Economic Times**

The title from a news article published during the depths of the recent global recession sums up the globalized nature of technology management: IT outsourcing deals, especially those leveraging offshore and globally distributed teams, continue to get larger and more intricate. Recent decades have seen the emergence and growth of multinational and transnational corporations that are comfortable operating in multiple geographies. Use of pervasive, cheap communication technologies — including Web collaboration software, e-mail, and other tools — is also accelerating the move toward globalization.

In the midst of this larger transition, the enterprise architect continues to evolve from an IT-centric function to a role organizations use to align IT with business. Enterprise architects, who constantly strive to drive strategic technology thinking in their organizations, are also recognizing the practical impact of sourcing on complex technology initiatives and programs. The enterprise architecture (EA) challenge in a sourcing context is twofold: on the one hand, to facilitate streamlined sourcing of key aspects of architecture definition and design, and on the other, to ensure continuity of organizational knowledge, processes, and intellectual property while leveraging external vendors and consultants.

In this chapter, we will examine some of the drivers, advantages, and challenges of global sourcing that enterprise architects can help address. The article will also provide some ideas on addressing challenges and leveraging external EA consultants, while offering some valuable insights and recommendations.

ENTERPRISE ARCHITECTS AND SOURCING

In a sourcing deal, the relationship between client and vendor organizations is detailed in service-level agreements (SLAs), operating-level agreements (OLAs), and individual statements of work (SOWs). These formal documents focus on delivery of key work tasks and work products by vendors. To ensure success of technology sourcing, enterprise architects are increasingly being expected to get involved in the business of sourcing. This includes defining RFPs, evaluating vendors, negotiating contracts, and defining SLAs and OLAs, while also providing technology oversight for strategic initiatives.

Leaders in the client organization take on vendor relationship management at different stages of the lifecycle, including contract negotiation, ongoing relationship management, and guaranteeing delivery of work products. Specialized vendor management teams define and negotiate contracts, with the assistance of line of business (LOB) leaders. LOB technology leaders generally manage the ongoing relationship with vendors, including handoff of specifications and review of vendors' work products. Enterprise architects provide inputs during key stages of the lifecycle.

Vendors are also expected to engage dedicated architects for large technology initiatives. These vendor architects are responsible for the architecting, designing, and delivery of solutions. Though some of these vendor architects take on tasks that would traditionally be the responsibility of inhouse solution architects, many organizations continue to employ inhouse solution architects as well. The intent of this "redundancy" is to make sure that checks and balances are maintained and the organization retains the depth it needs in technology areas. And while some of the solution architecture activities may be delegated to sourcing vendors, the enterprise architect role continues to be retained inhouse. In a recent survey of enterprise architects,[2]   only 1% of survey respondents indicated that they outsourced EA to strategic vendors. The survey authors concluded:

Enterprise Architecture is a critical function in managing the IT function. Therefore, it is difficult to see how an organization can give up this critical lever, unless it has decided to outsource IT as a whole (which also has not necessarily proven successful in most environments).[3]

Sourcing relationships enable IT stakeholders to delegate the architecting, design, and development of subsystems, or even entire systems. However, responsibility for seeing that technology solutions continually meet business requirements remains vested with the client's technology leaders and LOB stakeholders, including enterprise architects. In a sourcing context, enterprise architects take ownership of systems integration for large, complex programs and conduct ongoing reviews to ensure solutions delivered by vendors meet strategic objectives.

MANAGING AND ADDRESSING EA CHALLENGES IN A SOURCING CONTEXT" (CAPITALIZE EA)

Some architects with whom I have interacted at client organizations have expressed reservations about not being included in the business decisions involving sourcing strategies; others have successfully embraced sourcing. Organizations that have adopted sourcing strategies have also benefitted from enterprise architects who help in shaping some of the best practices. In a sourcing context, typical challenges that enterprise architects can address include:

- Loss of technical expertise due to sourcing
- The need to coordinate multivendor scenarios
- Vendors lacking knowledge of organizational dynamics
- Vendors lacking specific business context
- Vendors not up to date on organizational processes, acronyms, and jargon

In this section, I examine some of the common architectural challenges encountered in sourcing situations and highlight best practices for addressing them.

Loss of Technical Expertise Due to Sourcing

The presumption behind sourcing is that the vendor will, and generally does, bring in technical depth and hands-on expertise for IT projects and programs where they are engaged. Traditionally, inhouse architects, designers, and other senior technologists would be involved in complex technology initiatives. As they helped to solve problems — say, configuring networks, servers, and firewalls or defining key interfaces with external partners — enterprise architects would add back to the organizational knowledge ecosystem. This would also lead to communities of practice, and such organizational expertise could be tapped at any point in time.

With traditional sourcing, and colocated teams, there was a decoupling between ownership of core architectural aspects and the other tactical design and development work. Traditionally, architectural work would be managed by inhouse architects, while external contractors and vendors would supplement aspects of the software development and support lifecycles. With the advent of larger sourcing initiatives, such technical work — including architecture definition — is being sourced to vendors, who are expected to bring in technical depth and breadth along with an external perspective on problem solving. Offshoring adds additional layers of communication, since teams will not be geographically colocated.

Enterprise architects need to address the risk that sourcing deals will lead to a dilution of technical depth within the organization. The challenge is that outsourcing SLAs are generally designed to provide continuity of service independent of people. For example, Mr. X, the vendor architect who helped configure a complex

integration solution with a partner gateway, will probably move to another engagement for another client elsewhere in the world after the current project. Most SLAs will not bind the vendor to providing the services of Mr. X, but rather will call for the expertise from a person with skills similar to Mr. X's.

Addressing the Challenge: Retaining Organizational Technology Depth

Enterprise architects can take concrete steps and work with stakeholders to make certain that technology depth and knowledge continue to be retained in the organizational ecosystem. Specific ideas include:

- Decoupling the knowledge base from people. Organizations use several tools and techniques to ensure system knowledge is retained in the organizational ecosystem. This includes leveraging blogs, wikis, and other collaborative platforms. Effective search capability built into the tools means that knowledge will be available when required.

- Ensuring regular knowledge sharing. Enterprise architects need to take the lead in establishing regular interactions and brainstorming sessions among members of the technical community. Vendor technologists and architects should be included in the community of practice. Simple and inexpensive tools like e-mail lists, team wikis, and blogs are powerful ways of fostering communities of practice.

- Continuing to be involved in architecturally significant projects and programs. While it may not be practical or productive for enterprise architects to participate in all projects, being involved in technically challenging initiatives is a key success factor in achieving alignment with enterprise strategies.

Coordinating Multivendor Scenarios

Multisourcing strategies involve either working with multiple vendors for different aspects of larger programs or sourcing different functional areas to more than one vendor. Multisourcing enables enterprise IT organizations to increase efficiencies and improve service quality by leveraging the capabilities of multiple vendor partners. Organizations adopt multisourcing strategies as a risk mitigation approach and to leverage best-of-breed services.

Multisourcing requires two or more vendors to work together to deliver an end-to-end service for which no single vendor is fully accountable. Diffused accountability makes vendor coordination and collaboration difficult. For the enterprise architect, this presents the formidable challenge of ensuring a seamless handoff between vendors for different aspects of the end-state solution design.

Addressing the Challenge: Technology Management in Multivendor Scenarios

Best practices in managing multivendor scenarios include:

- Delegation of responsibilities. Enterprise architects should insist that vendors identify and delegate technical responsibilities to named vendor architects. This introduces accountability in that there is a single "throat to choke" if there is a problem. It also reduces layers of communication between and across vendor teams.

- Periodic communication. Periodic communication between stakeholders is a hallmark of well-run programs. Enterprise architects should take on the responsibility of clearing communication channels between and across technical communities — this includes communication between inhouse architects and technologists from the vendors' teams. Open communication enables issues and challenges to be identified early on.

- Reviewing vendor SLAs and SOWs for architecturally significant challenges.

See the "Enterprise Architects and Multisourcing" sidebar, which discusses how enterprise architects helped one organization tackle some multisourcing challenges.

Vendors Lack Knowledge of Organizational Dynamics

The success of larger technology programs and initiatives depends on several factors, including a good understanding of organizational dynamics. Organizations tend to evolve, and dynamics between people and processes at any given time will reflect collective experiences. Technology initiatives frequently fail because of inadequate consideration of the effects of the proposed changes on organizations. Sourcing vendors and consultants are expected to bring in ideas and external best practices, but they will almost certainly lack organizational knowledge and depth, especially in the softer aspects of organizational dynamics.

Effective enterprise architects gain the respect and trust of stakeholders by bridging the business-IT divide. Doing so requires a good understanding of organizational dynamics and constraints. An enterprise architect also contributes by being a repository of organizational memories, especially of past landmines and failures. Awareness of these pitfalls may not be described in project plans and formal documents, but it may be critical to project success.

Addressing the Challenge: Guiding Vendors in Organizational Dynamics

Details of organizational dynamics are hard to capture in formal vendor contracts (SLAs/OLAs), and it may be difficult to formally communicate these to vendors. Enterprise architects, vested in the success of organizational initiatives, should take the lead in guiding vendor teams through organizational dynamics, especially around challenges that could impact the technology landscape and solutions. Areas where an enterprise architect's guidance will add value include:

- Tool selection. Vendor teams may be charged with evaluating and selecting software tools and products. Selecting tools that can be reused across the enterprise may require aligning with a wider group of stakeholders. Enterprise architects should weigh in to reduce conflicts of interest where LOB stakeholders may want to pursue an independent tools or technology strategy.

- Defining success criteria for strategic engagements. Vendors may be engaged by LOB stakeholders to help in defining technology roadmaps and architectural strategies for their groups or divisions. Enterprise architects should provide strategic guidance to such initiatives, which may have long-term or strategic impacts beyond the LOBs.

Box 12.1

ENTERPRISE ARCHITECTS AND MULTISOURCING

I present here a real-world example of a multisourcing strategy being adopted by a global agribusiness company headquartered in Europe. Representing my employer, Infosys Technologies, I was engaged with enterprise architects of the organization. The focus of my engagement was to consult with the client on its enterprise architecture tools and modeling strategy.

The company is focused on two main types of products, seeds and crop protection, and has operations in over 90 countries. It had adopted a multisourcing strategy that defines a clear separation of responsibility between technology vendors. One vendor was responsible for production support and data center management. Another vendor was responsible for network management. Infosys

Box 12.1

teams along with teams from another IT service provider were involved in application development and maintenance. The portfolio consisted of bespoke applications, COTS products, along with instances of ERP (SAP) systems.

Ensuring communication, coordination, and delivery of programs and managing handoffs between vendor teams were the primary responsibilities of the client's program managers. During larger initiatives, enterprise architects would be involved in technology architecture and design reviews. There were more than a few instances in which large programs encountered challenges in the handoff between vendors. Enterprise architects worked with key stakeholders to conduct a root-cause analysis of the difficulties. It turned out there were several problems having to do with communication of technical details, description of system integration, and handoff of design descriptions, as well as disagreements about the right tools and techniques to use and the design patterns to be adopted. There was a distinct gap in communication both between vendor teams and between individual vendor teams and the client architects.

Enterprise architects began to get involved in reviewing SLAs and vendor SOWs to ensure that technical integration aspects were addressed adequately. The architects also began adopting some of the best practices I describe in the article. Vendors were asked to designate lead architects who would be accountable for key programs and essentially act as the single point of contact when it came to technical decisions. The enterprise architects started inviting vendor architects to key architectural strategy sessions and meetings and began to foster a culture of collaboration between key vendor technologists.

As the enterprise architects became proactively engaged

Box 12.1

CONTINUED...

in programs, they reduced the need for reactive fire-fighting. Their involvement in the sourcing lifecycle led to a noticeable reduction in sourcing program failures.

Vendors May Lack Subject Matter and Technology Base Expertise

Ideal sourcing strategies involve evaluating vendors on their technology depth and knowledge of business processes as they pertain to organizational and project needs. However, it may not be realistic to expect vendors' teams to mirror all dimensions of skills in your organization. These teams may lack subject matter expertise and familiarity with your organization's preexisting technology base, configurations, and process of interacting with business users.

Consider the example of a telecommunications program involving data interchange of access service request (ASR) forms in an ordering and billing forum (OBF) between your organization and another telco. The core technology platform may include DB2 databases and legacy CICS programs running on a mainframe and Java and XML technologies for external interfacing. While this is a typical scenario for many telecommuni- cations organizations, it may not be practical to expect an exact match of skills in all the technologies as well as OBF and ASRs when outsourcing the system to a vendor. The vendor (or vendors) may be able to engage architects with a background in the eclectic mix of technologies required or with some telecom experience — but probably not both. This is a challenge I have explored elsewhere.[4, 5]

Addressing the Challenge: Familiarizing Vendor Teams with Your Business Context

Enterprise architects should work with vendor archtects and specialists to familiarize them with the salient aspects of the organization's technology management lifecycle. This can be done by:

- Requiring the vendor to invest time and effort in orienting their teams and technologists in specific IS and key business processes.
- Facilitating interactions between vendors' teams and internal subject matter experts for additional areas of focus.

Vendors Not Up to Date on Organizational Processes, Acronyms, and Jargon

IT organizations, especially in larger enterprises, continually absorb best practices from the industry while adapting internal practices for specific needs. Most organizations evolve their distinct taxonomies, acronyms, and governance structures with unique roles, titles, and organizational structures. Organizations also develop specific IT/IS management hierarchies, groups, and project/program naming conventions and standards. In addition, teams may routinely interchange use of vendor, product, project, and group names.

Those who have spent any time in the organization are bound to forget the extent to which the acronyms drive conversations. All this can be complex even for those within the organization, to say nothing of vendor technologists.

Addressing the Challenge: Bringing Partners Up to Speed on Organizational Processes

As a consulting enterprise architect, I typically spend the first week of an engagement on two key areas: cut- ting through specific business context and challenges while also coming to grips with important jargon and three- and four-letter acronyms prevalent in the organizational vocabulary.

Some of the techniques and best practices for helping vendor architects come up to speed on all internal acronyms and naming standards include:

- Leveraging tools and technologies. Corporate intranets, wikis, and shared folders are simple and powerful platforms. To ensure ongoing viability, enterprise architects should set the expectation that architects who are coming on board will also be responsible for updating the content.
- Investing in orientation programs. Periodically scheduled orientation programs are an ideal way to introduce new members to architectural standards and practices. If face-to-face interactions are not possible, or are expensive, programs can be designed using online collaborative tools. The client and vendors should budget time and effort for such orientation programs when defining SLAs. In some instances, the client organization may have an ongoing orientation program that could be extended to include newer vendor team members at minimal cost.

CONCLUSION

Large-scale sourcing requires organizations to be pre- pared to follow through and deliver complex technology programs with vendors that may have geographically distributed teams. Enterprise architects have a key role to play in the success of such programs, especially in ensuring the design and delivery of the solutions vendors are contracted to provide. While it may not be practical or productive for enterprise architects to be involved in all projects, being involved in technically challenging initiatives is a key success factor in achieving alignment with enterprise strategies. Adopting and extending the best practices should help technologists address the challenges of sourcing and ensure alignment with business drivers and goals.

NOTES

1. Roy, Kumar Shankar. "Recession? TCS, Infosys, Wipro, HCL Bag Large Outsourcing Deals." Economic Times, 27 July 2009.

2. "Enterprise Architecture Expands its Role in Strategic Business Transformation (Infosys Enterprise Architecture Survey 2008/2009)." Infosys, 200[8?].

3. Infosys. See 2.

4. Babu K, Mohan. "Is IT-Business/Domain Knowledge Overrated?" Managing Offshore IT, Infosys, 21 February 2007 (http://infosysblogs.com/managing-offshore-it/2007/02/is_businessdomain_knwledge_ove_1.html).

5. Babu K, Mohan. "Is IT-Business/Domain Knowledge [While Offshoring] Overrated? ... Continued." Managing Offshore IT, Infosys, 11 May 2007 (http://infosysblogs.com/managing-off-shore-it/2007/05/is_itbusinessdomain_knowledge.html).

Offshoring Trends and Positioning

- ⚏ Position Statements on Offshoring
- ⚏ Offshoring Training Trends

POSITION STATEMENTS ON OFFSHORING

The trends in offshoring and outsourcing have not gone unnoticed by premier industry association and bodies across the globe. Indian software organizations have gone all out to promote India as a preferred outsourcing destination. For instance, the National Association of Software & Service Companies (NASSCOM) states[1] *"Leading global business intelligence and consultancy firms such as Giga, Forrester Research and McKinsey & Co. have cited various reasons for the increase of offshore outsourcing by MNCs to India."* Similarly, other industry bodies across the globe have taken positioning stance on offshoring to guide their professionals and members. The Institute of Electrical and Electronics Engineers (IEEE) in the US[2] takes a cautionary stance on offshoring stating, *"Whether the United States will benefit from the offshoring of jobs will ultimately depend on how the process is implemented. As in all competitions, there will be winners and losers."* Another national group in America, the Association for Computing Machinery[3] (ACM) in the US says, *"The great majority of*

ACM's members have been touched in some way by the practice of out-sourcing IT jobs. Council was quick to respond to this matter by creating the Job Migration Task Force, the purpose of which is to examine the ram-ifications of offshoring and to produce its findings in a white paper that will be widely disseminated and referenced." An extract from the British Computer Society's statements on Offshore Outsourcing follows.

APP. A.1

BCS[4] POSITION STATEMENTS: OFFSHORE OUTSOURCING

We are witnessing a major shift in the global IT services industry. Countries such as India, Ireland, Russia and South Africa are developing IT service industries and the offshore outsourcing market is growing rapidly. India currently holds around 80 % of the market. Ovum Holway has predicted that offshore sector revenues generated in the UK will more than double between 2003 and 2006 to reach over £1 billion. The total for Europe will hit £2 billion. Research organization IDC estimates that, in large companies, 60 to 80% of all IT contract negotiations now include an element of offshore working. Offshore outsourcing can offer significant benefits:

- IT services can be provided at substantially reduced prices. Offshore outsourcing companies in, say, India can afford to pay what are relatively high salary levels for well-qualified staff and still undercut companies carrying out work in the UK. IT jobs in parts of India attract some of the best graduates because they are well paid and provide an excellent working environment in custom-built business parks.
- Countries such as India and China have large numbers of well-educated professionals and can offer a highly trained, flexible workforce on demand. India produces 500,000 new English speaking IT and engineering graduates each year.

BCS⁴ Position Statements: Continued...

Jobs in the IT services sector are highly sought after and IT professionals are enthusiastic and committed.

- By combining services provided from different countries based in different time zones, global outsourcing suppliers can offer services 24 hours a day, 7 days a week throughout the year.
- Whilst undoubtedly there have been some concerns with service quality, many of the major offshore outsourcing companies have invested heavily in developing quality management processes and achieving accreditation to international standards such as ISO 9001. Over half of all organizations in the world that meet the USA Software Engineering Institute's Capability Maturity Model (CMM) Level 5 standard are based in India.

UK companies will want to exploit the advantages offered by offshore outsourcing services to maintain their competitiveness in the global marketplace. The British Computer Society promotes the exploitation of IT to deliver maximum business benefit and recognizes that attempts to regulate the market will ultimately harm British interests. There are, however, a number of factors that need to be taken into account if organizations are not to fall victim to the pitfalls of offshore outsourcing:

- Offshore outsourcing is not a panacea. It can be used successfully to develop applications that are well defined, with specifications that are unlikely to change during the course of development and require few links to other applications. Offshore outsourcing is unsuitable for innovative solutions that require prototyping and need to meet changing business requirements.

App. A.1

BCS[4] POSITION STATEMENTS: CONTINUED...

- Political instability in some parts of the world increases risk. Some organizations outsource to countries such as Canada or Ireland to avoid these problems although it is not possible to make such significant savings in these countries.
- To be successfully used, offshore outsourcing must be well managed by staff who not only have sound IT skills but also are able to develop effective supplier relationships, monitor performance and negotiate contract changes.
- The implementation of offshore outsourcing will sometimes lead to a loss of UK-based IT jobs. Organizations should take care not to lose skills that are vital to the ongoing management of outsourced services. In recent years, as the UK economy has slowed, IT skills have been more readily available than in the previous decade. As the economy improves, IT staff may again be in short supply.

The growth in offshore outsourcing is naturally of concern to IT professionals in the UK, especially now that the IT job market is at a low point. Ovum Holway has forecast that between 20,000 and 25,000 jobs may be lost in the UK IT industry over the next few years as a direct result of work moving offshore. The position is not entirely clear since some commentators have predicted a future substantial increase in demand for application developers as processors are integrated into many different products.

We anticipate an impact on UK IT salary levels, which rocketed to unsustainable levels in the late 1990s. In the past, even junior, inexperienced IT staff could command very high wage rates but we are unlikely to see a return to this situation. Salary levels in developing countries such as India may well rise as their IT service industries develop, but for many years to come there will be countries able to offer quality services at low costs.

App. A.1

BCS[4] POSITION STATEMENTS: CONTINUED...

The British Computer Society will be monitoring the impact of the growth in offshore outsourcing and reviewing a number of areas including:

- A renewed focus on quality management processes in UK-based IT organizations.
- Identifying and developing the skills needed by UK IT staff so that they can add value that cannot be matched by offshore workers.
- Promoting the competitive strengths of UK IT professionals, especially in leading edge, innovative technologies.
- Encouraging organizations and government to retrain and re-deploy any IT professionals who find themselves without work as a result of offshore outsourcing.

OFFSHORING TRAINING TRENDS

The demand for training and enablement of offshoring managers has lead to the emergence of an increasing number of 'Level 101' training and courses targeted towards the offshoring segment [Ref: Table A.1 below]. These courses attempt to strengthen the existing management training and focus on aspects of business management, program and project management along with an emphasis on the *softer* side including cross cultural communication and nuances of managing across time and space boundaries. In an opinion column titled *'Offshoring is not a Choice, it's a Reality,'* Satwik Seshasai[5] who helped start one of the first courses in Offshore Outsourcing at MIT says *'Education opportunities will be created where engineers are required to go (physically or virtually) to foreign countries and teach them about our culture, and learn about their culture as "ambassadors".'*

Course	Offshoring
Offshore Project Management Fundamentals	MGT275: Offered by: Washington University in St. Louis; http://www.cait.wustl.edu/courses/MGT275.co
"Managing Globally"	Carlson Business School's International Program; http://www.carlsonschool.umn.edu/Page615.aspx
Special Seminar in International Management—Offshoring	MIT Sloan School of Management; http://web.mit.edu/outsourcing/
Strategy, Leadership, and Governance: The Global Outsourcing of Key Business Processes	Harvard Business School; http://www.exed.hbs.edu/programs/slg/
Outsourcing	The University of Auckland Short Courses; http://www.shortcourses.auckland.ac.nz/index.cfm?fuseaction=course.detail&course_id=167&layout=default
Bentley Interactive Workshop	Bentley College; http://www.bentley.edu/news-events/pr_view.cfm?id=1609
MBA 590—Management of IT and BPO	University of Illinois, Chicago; http://www.uic.edu/cba/mba/programinfo/programoverview/ptsspring05.html

Table A.1   Indicative list of cources focusing on offshoring management

A sample curriculum of a course on 'Offshore Project Management Fundamentals[6], follows:

App. A.2

TREND: OFFSHORE PROJECT MANAGEMENT FUNDAMENTALS

(Course: MGT275[6] offered by: Washington University in St. Louis)

Description: Offshore IT outsourcing to India and other countries has moved from a curiosity to a key corporate strategy for many organizations. The information technology services industry is evolving at a fast pace, roles are changing dramatically and people in the midst of it are feeling the need for a set of skills! To manage a global project, the project managers will need to be experts on defining requirements, managing change, communications, cultural sensitivity, planning for and conducting project reviews and negotiations. In this changing business environment, the importance of good offshore project management is just beginning to be recognized and addressed. It is important to maintain a sophisticated project and program management function addressing transition management, governance, performance management and quality management. In this workshop, we look at some of the challenges and best practices in the area of offshore project management in order to help IT managers keep up-to-date on this rapidly evolving way of working. This workshop is for IT managers who would like to progress to higher levels of understanding the finer aspects of working with offshore teams. In addition to gaining key insights, this workshop will give participants the opportunity to meet and share learning with those at other corporations who are facing similar challenges. Workshop content will include presentations and discussions of facts, theory and case studies. There will be featured speakers from the industry who will also be available for discussions. Also noteworthy is that the workshop will give participants a taste of Indian culture, both figuratively and literally, with local cuisine

App. A.2

TREND: CONTINUED...

Audience: This workshop will be valuable to a wide range of professionals, from those currently engaged in managing offshore outsourced projects to those simply contemplating it.

Prerequisites: There is no prerequisite for this course.

Objectives: After completing this seminar, participants will be able to:

- Realize that offshore project management is about Challenge, Leadership & Achievement
- Understand the enhanced roles and responsibilities for successful offshore project management
- Build internal capabilities for the successful project management of offshore initiatives
- Assess the requirements for successful offshore project management in their own organizations
- Know the best business practices for project management in offshore outsourcing

Course Outline:

Introduction

Explore the Experience—with the Facilitator

Explore the Experience—within You (Assignment)

Explore the Experience—Challenges

The Prima of Offshoring—Benefits, Risks & Impact

Group Discussion—Managing change to Sustain Growth

Offshore Project Management—The Difference

Offshore Project Management—The Transition

Group Discussion—SWOT for Offshore Sourcing

Critical Success Factors for Offshore Project Management

Simplifying the Complexity: 4R's

    Risk

    Relationship

App. A.2

TREND: CONTINUED...

    Resources
    Reward
   Simplifying the Complexity: 4C's
    Cost & Commercial
    Control
    Communication
    Culture
   Value Reinforcement
   Explore the Experience—Benefits
   WAY FORWARD
   Case Studies
   The Handbook
Elective Hours (Cum. Classroom Time): 12 hours

NOTES:

1. 'Why India' [NASSCOM: http://www.nasscom.org/artdisplay.asp?cat_id=28]

2. IEEE USA: [http://www.ieeeusa.org/policy/positions/offshoring.asp]

3. Communications of the ACM [Volume 47, Number 12 (2004)]

4. British Computer Society [Refer http://www.bcs.org/BCS/News/ PositionsAndResponses/Positions/offshore/offshore.htm]

5 Washington University in St. Louis [Source: Washington University in St. Louis; http://www.cait.wustl.edu/courses/MGT275.co]

6. Offshoring is not a Choice, it's a Reality [Satwik Seshasai, MIT Alumni Newsletter http://alum.mit.edu/ne/whatmatters/200406/ index.html]

APPENDIX B

Offshoring Models

- 🖥 Nearshoring
- 🖥 Rural Sourcing
- 🖥 Homeshoring
- 🖥 Multishoring

The growing awareness of the working models and strategies of outsourcing application development and maintenance has brought forth intense rivalry among countries, especially from the underdeveloped economies that have realized that they can get the most out of their human capital to compete in the global marketplace. *"On almost daily basis companies announce offshoring-decisions in order to cut costs. The consequence is a massive outsourcing of highly-skilled white-collar jobs such as finance analysts or software specialists into low-wage countries. Many economists advocate this form of international division of labor, since a global competition would naturally call for a global solution."* says Julia Trampel[1]. Offshoring, the practice of sourcing technology work to low-wage countries, is not the only outsourcing model prevalent in the IT industry, even though offshore outsourcing is almost synonymous with outsourcing. Companies and IT vendors around the globe are actively touting alternative sourcing models including:

1. NEARSHORING

Wikipedia[2] defines it as, *"Nearshoring is a form of Outsourcing in which business processes are relocated to cheaper, still, geographically closer locations. Nearshoring can also be contrasted with Offshoring or OffshoreOutsourcing, which implies relocation of business processes to cheaper, typically foreign locations."* This trend is predominant in North America and Europe with US companies sourcing work to Canada and Mexico, and European countries outsourcing work to East European nations. According to Andreas Floth[3] of PA Consulting, *"IT services businesses in Eastern Europe hope that EU accession will boost cross-border trade, in particular with Germany. Indian and Asian service providers have found it harder to make headway in northern and central Europe for cultural as much as technical reasons. Far fewer Indian graduates speak German than English, and there is less experience of the culture and business of mainland Europe"* Interestingly, even Indian firms have been exploring nearshoring options to work towards truly global sourcing models.

2. RURAL SOURCING

Several startups in the US and other nations are exploring business models around 'Rural Sourcing', essentially outsourcing domestically to lower-cost rural regions like Arkansas and New Mexico among other states. For instance, a company called Rural Sourcing[4] claims to offer services such as application maintenance and Internet development for roughly 40% less than what other domestic tech outsourcers charge. Although the scale of operations and growth potential is doubtful, such outsourcing vendors are getting some press mention based on their 'anti offshoring' pitch. In a recent interview[5] the CEO of Ciber, a Colorado based company, Mac Slingerlend was quoted as saying, *"There are many American labor markets outside the traditional technology centers that have skilled but under-utilized IT (information technology) workers who can get IT projects done faster and cheaper."*

3.   HOMESHORING

Among the innovative trends in the outsourcing space is the move by a few enterprising companies to 'mop up' offshoring deals that have gone sour. In an interview[6] the executives of Decision Design claimed that, *"the notion of 'homeshoring centers' in the United States to offer low costs to customers. In part by locating offices on the fringe of Silicon Valley and Chicago, the company claims that it can deliver savings of 30% to 60% below typical onshore development costs.'* Interestingly, the research firm IDC[7] used the term to describe call center workers handling calls from their homes. According to IDC, the so-called homeshoring or homesourcing in certain situations can boost productivity while cutting costs.

4.   MULTISHORING

Several organizations have begun to outsource work to multiple offshore locations, sometimes deciding based on the job to be done, the availability of skills and other infrastructural aspects. This strategy is most beneficial in operational scenarios where companies need to bridge the time zone differences and to utilize a '24 X 7' work-day model with the least incremental cost. The caveat of adopting this model is the management and administrative overheads that can come into play.

The list of terms presented above are an indication of terms being coined by offshoring consultants and vendors. Some of the terms exist merely to differentiate a vendor or consultant's sourcing strategies, while a few others have been coined to merely get some PR mileage and to create a *buzz* in the marketplace. Most of these sourcing strategies have a common thread running through: they are variants of the popular offshore outsourcing models with the key differentiator being the location where the work is done.

NOTES

1. To Offshore or Nearshore IT services?—An investigation using transaction cost theory [Julia Trampel]

2. Wikipedia; the free encyclopedia [http://en.wikipedia.org/wiki]

3. News Release: PA Consulting Group says 'Nearshoring' in Eastern Europe is a Real Alternative—The European Outsourcing Summit— June 2004 [PA Consulting Group, Wednesday 23 June 2004]

4. Rural Sourcing Inc. [http://www.ruralsource.com/]

5. Colorado firm outsources to Oklahoma [Cnet News.com, January 16, 2005]

6. Indian IT firms to set up 'nearshore' bases in Europe [*silicon.com*, by Andy McCue, October 11 2004]

7. U.S. firm sweet on offshoring deals gone sour [Ed Frauenheim, CNET News.com, January 10, 2005]

8. 'Homeshoring' to trump offshoring? [Ed Frauenheim, CNET News.com, December 21, 2004]

INDEX

Viligos   *6.13*

re-engineering   *5.14; 7.1; 7.4*

validation services   *5.15*

R&D   *3.3; 1.14; 5.15*

Zensar   *3.3*

IT Strategy   *3.12; 5.13*

Mega trend   *1.1; 1.3*

Gregory Millman   *1.1*

ERP   *1.2; 5.13*

Prahalad C.K   *1.23; 9.6*

EAI   *5.13*

CRM   *1.2; 6.16; 9.5*

BPM   *5.13*

COTS   *5.13*

Business Intellegence   *5.13*

content management   *5.13*

data wearhousing   *5.13*

Data Mining   *5.13*

Teams

   Cross functional teams   *1.2*

   Self managed teams   *1.2*

   Geographically distributed teams   *9.1*

   Virtual teams   *9.1*

   Managing Global Teams   *9.2*

   Cross cultural teams   *9.7*

   Team Dynamics   *9.9*

   Team Culture   *9.13*

Body Shops   *1.4*

Offshoring   *1.1*

   Offshoring definitions   *1.1*

   Outsourcing Strategy   *1.12; 2.1; 3.3; 3.8*

   Waterfall model of Offshoring Strategy   *2.2*

   Local Sourcing   *1.5*

   Mixed Sourcing   *1.6*

   Global Delivery Model   *1.7*

   Strategic Outsourcing   *1.8*

   Joint Ventures   *2.8*

   Captive Development Centre   *2.8*

   Sourcing to vendors   *2.8; 2.13*

   Build-Operate-Transfer (BOT)   *2.1*

   Build-Own-Operate-Transfer (BOOT)   *2.1*

   Subsidiaries   *2.11*

   Offshore Development Centre (ODC)   *2.15*

   Captive Development Centre   *2.11*

   Global Delivery Model   *2.15*

   Multi-vendor Offshoring   *2.15*

ITO and BPO   *1.8*

Cutter IT Journal *1.8*

Stern Stewart & Co *1.8*

Innovative Vs Sustain *1.1*

Line of Business (LOB) *1.1*

Return on Investment (ROI)
1.1; 4.13; 6.2; 5.13; 3.12; 7.4

Core Competence *1.11*

Risk

    Risk awareness *1.11*

    Risk Analysis *2.4*

    Risk Mitigation *2.1*

    Organisational Risk *2.2*

    Technical Risk *2.2*

    External Risk *2.2*

    New Market Risk *2.2*

    Risk Tolerance *2.27*

Ian Hayes *1.12; 1.14*

Service Level Agreements
(SLA) *7.6; 8.11; 7.6; 7.10; 3.10;
3.8; 3.11; 1.13*

Benchmarking *1.13; 3.9; 3.19*

Sweet Spot *1.14*

Mark Kobayashi-Hillary *1.15*

Washington University *1.21*

Offshoreability *2.1*

Piloting *2.4*

Outsourcing *2.4*

    Offshore/Onsite Mix *2.26*

    Offshoring study mission' in
    India *3.1*

    Offshoring Management
    Framework(OMF) *3.2; 3.5*

    Offshoring experts *3.3*

    Offshoring Transition *3.12*

    Offshoring Application De-
    velopment *6.14*

Information Technology Outsourcing *3.6*

Knowledge Management *2.6;
2.18; 5.13*

Programme Management *2.7*

2RentACoder.com *2.16*

Ralph Kilem *2.2*

Steve McConnell *2.22*

General Electric *3.3*

Cisco *3.3; 4.18*

Nokia *4.18*

Aberdeen Group *4.18*

Gartner *4.18*

Giga *4.18*

McKinsey's Global Institute
4.18

Lucent *4.18*

British Telecom *4.18*

Donald Knuth *3.6*

Deloitte report *3.7*

Service Level Agreement *3.7;
3.11*

Program Management *3.7*

Program Management Office
3.15; 8.12

Transition Management *3.7*

GE's *70:70:70* *3.9*

Thomas Lynch *3.11*

Contract Administration *3.17*

Legally Binding Contracts *2.9*

Change Management *3.18; 5.7;
7.16; 7.15; 8.7*

Infrastructure Management
3.18

Application Development &

Maintenance(ADM) *4.1*

Management Layer *4.2*

PMBOK *4.3; 4.10; 8.1 ; 5.1*

PMI *4.3; 8.22*

Onsite activities *4.3*

Offshore activities *4.4*

Global Project Management *4.7*

General Body of knowledge *4.9*

Rate of Interest (ROI) *4.13*

COCOMO *4.13*

Functional Point (FP) *4.13*

Use Case Points (UCP) *4.13*

   Infosys *1.7; 3.3; 8.18*

   Infosys Tools and processes *4.15*

Infosys Case *4.17*

Global IT Manager *4.19*

Erran Carmel *9.1; 4.19*

Project Execution *5.2; 5.12*

Controlling and Monitoring *5.4*

Source Offsite *5.8*

ISO *5.11*

CMM *5.11*

SEI *5.11*

Wipro *1.7; 2.24; 3.3; 5.11; 5.14; 7.12 ; 8.18*

System Integration *5.14*

Centres of Excellence (COE) *5.15*

Marc Andreesen *5.15*

Execution *6.1*

Rapid Application

Development (RAD) *6.3*

Extreme Programming (XP) *6.3*

IEEE *6.3*

Steve MCConnell *6.3*

Software Development Life Cycle (SDLC) *6.4; 6.8*

Application Life Cycle *7.2*

Maintenance Life Cycle *7.6*

Subject Matter Experts (SME) *6.5*

Request for Proposals (RFP) *6.5*

Request for Information (RFI) *6.5*

Statement of Work (SOW) *6.5*

Requirement Analysis (RA) *6.5*

Joint Analysis and Design (JAD) *6.5*

Proof of Concept (POC) *6.6*

Total Cost of Ownership (TCO) *6.7; 5.13*

Quality Assurance *4.14*

Testing *4.14; 5.8; 5.14; 6.9; 7.13; 9.5*

   Test Plan *6.11*

   Unit Test *6.8*

   System Integration Test (SIT) *6.9*

   User Accepting Testing (UAT) *6.9*

Standish Group *6.1*

Coding standards *6.11*

Vigilos *6.13*

Burton Swanson *7.1*

Andre Nadeau *7.1*

CGI Group *7.1*

Mayuram S Krishnan *7.2*

Line of Business (LOB) *7.6*

Change Control Board (CCB)
7.9

Maintenance extension to OMF
7.1

Tibco *5.15*

IBM *5.15; 1.15; 2.12*

Oracle *1.15; 2.12; 9.19*

Microsoft *1.14; 1.15; 2.12; 2.19;
5.15; 9.16; 9.19*

Microsoft Network *7.14; 8.13*

Microsoft Project *10.5; 8.18;
4.13*

Yahoo *8.13*

AOL *8.13*

ACM *7.15*

Configuration Management
7.15; 6.12

Peter Drucker *8.1; 9.21*

Hall, E.T. *8.2*

Hofstede, G *8.2*

Communication layer *8.5; 8.6*

communication management
plan *8.7*

Technologies of
Communication
    E-Mails *8.9*
    Instant Messengers *8.12*
    Blogging and Wikis *8.13*
    Video and Audio communi-
    cation *8.16*

Securities and Exchange
Commission (SEC) *8.11*

Harvard Business Review *8.14*

Wilcox Development Solutions
8.14

VOIP *8.16*

John Tuman *8.17*

David Pells *8.19*

GaramChai.com *8.21; 9.7*

Benjamin Limbach *9.8*

B-Players *9.14*

SMEs *9.15*

Frederick Herzberg *9.18*

Abraham Maslow *9.18*

Egoboo *9.19*

Gold Collar workers *9.20; 9.21*

Software Culture *9.1; 9.10; 9.16*

Knowledge
    Knowledge Sharing culture
    9.15
    Knowledge Management
    10.8

Strategic Inflexion Point *10.1*

Ed Sullivan *10.3*

Geffory Moore *10.6*

Bhaskar Chakravorti *10.6*

Technology vendors *10.13*

George A Steiner *10.15*

Digital Security *10.17*

Harvard Business Review
1.22; 8.14; 9.25 10.17

Nancy Mead *10.18*

Quality *5.1*